The NSTA Quick-Reference Guide to the NGSS

K–12

The NSTA Quick-Reference Guide to the NGSS

K–12

Edited by Ted Willard

Claire Reinburg, Director
Wendy Rubin, Managing Editor
Andrew Cooke, Senior Editor
Amanda O'Brien, Associate Editor
Amy America, Book Acquisitions Coordinator

Art and Design
Will Thomas Jr., Director
Cover design by Gina Toole Saunders

Printing and Production
Colton Gigot, Senior Production Manager

National Science Teaching Association
Erika Shugart, Executive Director
Cathy Iammartino, Publisher

405 E Laburnum Ave Ste 3, Richmond, VA 23222
www.nsta.org/store
For customer service inquiries, please call 800-277-5300.

22 21 20 19 9 8 7

Library of Congress Cataloging-in-Publication Data
Willard, Ted.
The NSTA quick-reference guide to the NGSS, K–12 / edited by Ted Willard.
pages cm
ISBN 978-1-941316-10-8—ISBN 978-1-941316-90-0 (electronic)
1. Science—Study and teaching (Elementary)—Standards—United States. 2. Science—Study and teaching (Secondary)—Standards--United States. I. National Science Teaching Association. II. Title.
LB1585.3.W595 2014
507.1—dc23
2014033833
Cataloging-in-Publication Data for the e-book are also available from the Library of Congress.

CONTENTS

CONTENTS

INTRODUCTION

Since the release of the first draft of the *Next Generation Science Standards* (*NGSS*), NSTA has been at the forefront in promoting the standards and helping science educators become familiar with and learn to navigate this exciting but complex document. When the final version was released and states began adopting the standards, NSTA started to develop resources that would assist educators with their implementation, including web seminars, virtual conferences, sessions and forums at conferences, books, and the NGSS@NSTA Hub (*http://ngss.nsta.org*)—a digital destination focusing on all things *NGSS*.

Along the way, NSTA learned that even the simplest of resources, such as a one-page cheat sheet, can be extremely useful. Many of those tools are collected in this volume, including

- a two-page cheat sheet that describes the practices, core ideas, and crosscutting concepts that make up the three dimensions described in *A Framework for K–12 Science Education;*
- an "Inside the *NGSS* Box" graphic that explains all of the individual sections of text that appear on a page of the *NGSS;*
- a Venn diagram comparing the practices in *NGSS* and *Common Core State Standards* in English language arts and mathematics; and
- matrixes showing how the *NGSS* are organized by topic and by disciplinary core idea.

We've also produced tables to describe the various parts of the standards. For example, the performance expectations describe what every student should know and be able to do by the end of a particular grade or grade span. These expectations are designed to assess the material in the foundation box, which includes

- science and engineering practices;
- disciplinary core ideas;
- crosscutting concepts;
- connections to engineering, technology, and applications of science; and
- connections to nature of science.

While summative assessments are required to focus on a particular combination of these components, curriculum developers and classroom teachers have the freedom to mix and match these components in a wide variety of ways. In fact, to learn any particular disciplinary core idea or crosscutting concept, students will need to engage in *multiple* practices in a well-thought-out sequence of learning experiences. The matrixes we developed and include in this book will help educators in their planning. There are two different sets of matrixes. The first set shows the K–12 progression of each of the components in the foundation box (e.g., practices, core ideas, or crosscutting concepts). These matrixes will help you understand how what students are expected to know and do in each grade span builds on what they have learned in earlier grades and prepares them for what they are expected to learn in later grades.

The second set of matrixes combines all the materials for a particular grade level together. For example, one of the matrixes focuses only on the science and engineering practices that students need to master in grades K–2.

The materials in this book should be a useful companion to the *NGSS*. The educators we have shared them with have found it helpful to photocopy particular pages for participants to use in a workshop or for colleagues to use during planning time.

ACKNOWLEDGMENTS

Production of *A Framework for K–12 Science Education* and the *Next Generation Science Standards* involved the work and contributions of thousands of educators, and I thank them for their efforts. Almost every word in this publication is drawn directly from those two documents, but any errors that appear here are mine. In addition I want to thank those educators involved in developing the documents that preceded *NGSS,* including the *Atlas of Science Literacy, National Science Education Standards, Benchmarks for Science Literacy,* and *Science for All Americans*. Finally, I thank the many educators working to make the vision of science literacy for all a reality for their students. This book is for you.

CHAPTER 1

Basics of *NGSS*

Three Dimensions of the *Next Generation Science Standards* (*NGSS*)

Science and Engineering Practices

Asking Questions and Defining Problems
A practice of science is to ask and refine questions that lead to descriptions and explanations of how the natural and designed world works and which can be empirically tested.

Engineering questions clarify problems to determine criteria for successful solutions and identify constraints to solve problems about the designed world.

Both scientists and engineers also ask questions to clarify the ideas of others.

Planning and Carrying Out Investigations
Scientists and engineers plan and carry out investigations in the field or laboratory, working collaboratively as well as individually. Their investigations are systematic and require clarifying what counts as data and identifying variables or parameters.

Engineering investigations identify the effectiveness, efficiency, and durability of designs under different conditions.

Analyzing and Interpreting Data
Scientific investigations produce data that must be analyzed in order to derive meaning. Because data patterns and trends are not always obvious, scientists use a range of tools—including tabulation, graphical interpretation, visualization, and statistical analysis—to identify the significant features and patterns in the data. Scientists identify sources of error in the investigations and calculate the degree of certainty in the results. Modern technology makes the collection of large data sets much easier, providing secondary sources for analysis.

Engineering investigations include analysis of data collected in the tests of designs. This allows comparison of different solutions and determines how well each meets specific design criteria—that is, which design best solves the problem within given constraints. Like scientists, engineers require a range of tools to identify patterns within data and interpret the results. Advances in science make analysis of proposed solutions more efficient and effective.

Developing and Using Models
A practice of both science and engineering is to use and construct models as helpful tools for representing ideas and explanations. These tools include diagrams, drawings, physical replicas, mathematical representations, analogies, and computer simulations.

Modeling tools are used to develop questions, predictions, and explanations; analyze and identify flaws in systems; and communicate ideas. Models are used to build and revise scientific explanations and proposed engineered systems. Measurements and observations are used to revise models and designs.

Constructing Explanations and Designing Solutions
The products of science are explanations and the products of engineering are solutions.

The goal of science is the construction of theories that provide explanatory accounts of the world. A theory becomes accepted when it has multiple lines of empirical evidence and greater explanatory power of phenomena than previous theories.

The goal of engineering design is to find a systematic solution to problems that is based on scientific knowledge and models of the material world. Each proposed solution results from a process of balancing competing criteria of desired functions, technical feasibility, cost, safety, aesthetics, and compliance with legal requirements. The optimal choice depends on how well the proposed solutions meet criteria and constraints.

Engaging in Argument From Evidence
Argumentation is the process by which explanations and solutions are reached.

In science and engineering, reasoning and argument based on evidence are essential to identifying the best explanation for a natural phenomenon or the best solution to a design problem. Scientists and engineers use argumentation to listen to, compare, and evaluate competing ideas and methods based on merits.

Scientists and engineers engage in argumentation when investigating a phenomenon, testing a design solution, resolving questions about measurements, building data models, and using evidence to identify strengths and weaknesses of claims.

Using Mathematics and Computational Thinking
In both science and engineering, mathematics and computation are fundamental tools for representing physical variables and their relationships. They are used for a range of tasks such as constructing simulations; statistically analyzing data; and recognizing, expressing, and applying quantitative relationships.

Mathematical and computational approaches enable scientists and engineers to predict the behavior of systems and test the validity of such predictions. Statistical methods are frequently used to identify significant patterns and establish correlational relationships.

Obtaining, Evaluating, and Communicating Information
Scientists and engineers must be able to communicate clearly and persuasively the ideas and methods they generate. Critiquing and communicating ideas individually and in groups is a critical professional activity.

Communicating information and ideas can be done in multiple ways: using tables, diagrams, graphs, models, and equations as well as orally, in writing, and through extended discussions. Scientists and engineers employ multiple sources to acquire information that is used to evaluate the merit and validity of claims, methods, and designs.

Disciplinary Core Ideas in Physical Science	Disciplinary Core Ideas in Life Science	Disciplinary Core Ideas in Earth and Space Science	Disciplinary Core Ideas in Engineering, Technology, and Applications of Science
PS1: Matter and Its Interactions PS1.A: Structure and Properties of Matter PS1.B: Chemical Reactions PS1.C: Nuclear Processes **PS2: Motion and Stability: Forces and Interactions** PS2.A: Forces and Motion PS2.B: Types of Interactions PS2.C: Stability and Instability in Physical Systems **PS3: Energy** PS3.A: Definitions of Energy PS3.B: Conservation of Energy and Energy Transfer PS3.C: Relationship Between Energy and Forces PS3.D: Energy in Chemical Processes and Everyday Life **PS4: Waves and Their Applications in Technologies for Information Transfer** PS4.A: Wave Properties PS4.B: Electromagnetic Radiation PS4.C: Information Technologies and Instrumentation	**LS1: From Molecules to Organisms: Structures and Processes** LS1.A: Structure and Function LS1.B: Growth and Development of Organisms LS1.C: Organization for Matter and Energy Flow in Organisms LS1.D: Information Processing **LS2: Ecosystems: Interactions, Energy, and Dynamics** LS2.A: Interdependent Relationships in Ecosystems LS2.B: Cycles of Matter and Energy Transfer in Ecosystems LS2.C: Ecosystem Dynamics, Functioning, and Resilience LS2.D: Social Interactions and Group Behavior **LS3: Heredity: Inheritance and Variation of Traits** LS3.A: Inheritance of Traits LS3.B: Variation of Traits **LS4: Biological Evolution: Unity and Diversity** LS4.A: Evidence of Common Ancestry and Diversity LS4.B: Natural Selection LS4.C: Adaptation LS4.D: Biodiversity and Humans	**ESS1: Earth's Place in the Universe** ESS1.A: The Universe and Its Stars ESS1.B: Earth and the Solar System ESS1.C: The History of Planet Earth **ESS2: Earth's Systems** ESS2.A: Earth Materials and Systems ESS2.B: Plate Tectonics and Large-Scale System Interactions ESS2.C: The Roles of Water in Earth's Surface Processes ESS2.D: Weather and Climate ESS2.E: Biogeology **ESS3: Earth and Human Activity** ESS3.A: Natural Resources ESS3.B: Natural Hazards ESS3.C: Human Impacts on Earth Systems ESS3.D: Global Climate Change	**ETS1: Engineering Design** ETS1.A: Defining and Delimiting an Engineering Problem ETS1.B: Developing Possible Solutions ETS1.C: Optimizing the Design Solution **ETS2: Links Among Engineering, Technology, Science, and Society** ETS2.A: Interdependence of Science, Engineering, and Technology ETS2.B: Influence of Engineering, Technology, and Science on Society and the Natural World

Crosscutting Concepts

Patterns
Observed patterns of forms and events guide organization and classification, and they prompt questions about relationships and the factors that influence them.

Cause and Effect: Mechanism and Prediction
Events have causes, sometimes simple, sometimes multifaceted. A major activity of science is investigating and explaining causal relationships and the mechanisms by which they are mediated. Such mechanisms can then be tested across given contexts and used to predict and explain events in new contexts.

Scale, Proportion, and Quantity
In considering phenomena, it is critical to recognize what is relevant at different measures of size, time, and energy and to recognize how changes in scale, proportion, or quantity affect a system's structure or performance.

Systems and System Models
Defining the system under study—specifying its boundaries and making explicit a model of that system—provides tools for understanding and testing ideas that are applicable throughout science and engineering.

Energy and Matter: Flows, Cycles, and Conservation
Tracking fluxes of energy and matter into, out of, and within systems helps one understand the systems' possibilities and limitations.

Structure and Function
The way in which an object or living thing is shaped and its substructure determine many of its properties and functions.

Stability and Change
For natural and built systems alike, conditions of stability and determinants of rates of change or evolution of a system are critical elements of study.

Chapter 1

Science and Engineering Practices*

Science and Engineering Practice 1: Asking Questions and Defining Problems

Questions are the engine that drive science and engineering.

Science asks

- What exists and what happens?
- Why does it happen?
- How does one know?

Engineering asks

- What can be done to address a particular human need or want?
- How can the need be better specified?
- What tools and technologies are available, or could be developed, for addressing this need?

Both science and engineering ask

- How does one communicate about phenomena, evidence, explanations, and design solutions?

Asking questions is essential to developing scientific habits of mind. Even for individuals who do not become scientists or engineers, the ability to ask well-defined questions is an important component of science literacy, helping to make them critical consumers of scientific knowledge.

Scientific questions arise in a variety of ways. They can be driven by curiosity about the world (e.g., Why is the sky blue?). They can be inspired by a model's or theory's predictions or by attempts to extend or refine a model or theory (e.g., How does the particle model of matter explain the incompressibility of liquids?). Or they can result from the need to provide better solutions to a problem. For example, the question of why it is impossible to siphon water above a height of 32 feet led Evangelista Torricelli (17th-century inventor of the barometer) to his discoveries about the atmosphere and the identification of a vacuum.

Questions are also important in engineering. Engineers must be able to ask probing questions in order to define an engineering problem. For example, they may ask: What is the need or desire that underlies the problem? What are the criteria (specifications) for a successful solution? What are the constraints? Other questions arise when generating possible solutions: Will this solution meet the design criteria? Can two or more ideas be combined to produce a better solution? What are the possible trade-offs? And more

---

* Excerpted with permission from: National Research Council (NRC). 2012. *A framework for K–12 science education: Practices, crosscutting concepts, and core ideas.* Washington, DC: National Academies Press.

questions arise when testing solutions: Which ideas should be tested? What evidence is needed to show which idea is optimal under the given constraints?

The experience of learning science and engineering should therefore develop students' ability to ask—and indeed, encourage them to ask—well-formulated questions that can be investigated empirically. Students also need to recognize the distinction between questions that can be answered empirically and those that are answerable only in other domains of knowledge or human experience.

GOALS

By grade 12, students should be able to

- Ask questions about the natural and human-built worlds—for example: Why are there seasons? What do bees do? Why did that structure collapse? How is electric power generated?
- Distinguish a scientific question (e.g., Why do helium balloons rise?) from a nonscientific question (Which of these colored balloons is the prettiest?).
- Formulate and refine questions that can be answered empirically in a science classroom and use them to design an inquiry or construct a pragmatic solution.
- Ask probing questions that seek to identify the premises of an argument, request further elaboration, refine a research question or engineering problem, or challenge the interpretation of a data set—for example: How do you know? What evidence supports that argument?
- Note features, patterns, or contradictions in observations and ask questions about them.
- For engineering, ask questions about the need or desire to be met in order to define constraints and specifications for a solution.

PROGRESSION

Students at any grade level should be able to ask questions of each other about the texts they read, the features of the phenomena they observe, and the conclusions they draw from their models or scientific investigations. For engineering, they should ask questions to define the problem to be solved and to elicit ideas that lead to the constraints and specifications for its solution. As they progress across the grades, their questions should become more relevant, focused, and sophisticated. Facilitating such evolution will require a classroom culture that respects and values good questions, that offers students opportunities to refine their questions and questioning strategies, and that incorporates the teaching of effective questioning strategies across all grade levels. As a result, students will become increasingly proficient at posing questions that request relevant empirical evidence; that seek to refine a model, an explanation, or an engineering problem; or that challenge the premise of an argument or the suitability of a design.

Science and Engineering Practice 2: Developing and Using Models

Scientists construct mental and conceptual models of phenomena. Mental models are internal, personal, idiosyncratic, incomplete, unstable, and essentially functional. They serve the purpose of being a tool for thinking with, making predictions, and making sense of experience. Conceptual models, the focus of this section, are, in contrast, explicit representations that are in some ways analogous to the phenomena they represent. Conceptual models allow scientists and engineers to better visualize and understand a phenomenon under investigation or develop a possible solution to a design problem. Used in science and engineering as either structural, functional, or behavioral analogs, albeit simplified, conceptual models include diagrams, physical replicas, mathematical representations, analogies, and computer simulations. Although they do not correspond exactly to the more complicated entity being modeled, they do bring certain features into focus while minimizing or obscuring others. Because all models contain approximations and assumptions that limit the range of validity of their application and the precision of their predictive power, it is important to recognize their limitations.

Conceptual models are in some senses the external articulation of the mental models that scientists hold and are strongly interrelated with mental models. Building an understanding of models and their role in science helps students to construct and revise mental models of phenomena. Better mental models, in turn, lead to a deeper understanding of science and enhanced scientific reasoning.

Scientists use models (from here on, for the sake of simplicity, we use the term "models" to refer to conceptual models rather than mental models) to represent their current understanding of a system (or parts of a system) under study, to aid in the development of questions and explanations, and to communicate ideas to others. Some of the models used by scientists are mathematical; for example, the ideal gas law is an equation derived from the model of a gas as a set of point masses engaged in perfectly elastic collisions with each other and the walls of the container—which is a simplified model based on the atomic theory of matter. For more complex systems, mathematical representations of physical systems are used to create computer simulations, which enable scientists to predict the behavior of otherwise intractable systems—for example, the effects of increasing atmospheric levels of carbon dioxide on agriculture in different regions of the world. Models can be evaluated and refined through an iterative cycle of comparing their predictions with the real world and then adjusting them, thereby potentially yielding insights into the phenomenon being modeled.

Engineering makes use of models to analyze existing systems; this allows engineers to see where or under what conditions flaws might develop or to test possible solutions to a new problem. Engineers also use models to visualize a design and take it to a higher level of refinement, to communicate a design's features to others, and as prototypes for testing design performance. Models, particularly modern computer simulations that encode relevant physical laws and properties of materials, can be especially helpful both in realizing and testing designs for structures, such as buildings, bridges, or aircraft, that are expensive to construct and that must survive extreme conditions that occur only on rare occasions. Other types of engineering problems also benefit from use of specialized computer-based simulations in their design and testing phases. But as in science, engineers who use models must be aware of their intrinsic limitations and test them against known situations to ensure that they are reliable.

GOALS

By grade 12, students should be able to

- Construct drawings or diagrams as representations of events or systems—for example, draw a picture of an insect with labeled features, represent what happens to the water in a puddle as it is warmed by the Sun, or represent a simple physical model of a real-world object and use it as the basis of an explanation or to make predictions about how the system will behave in specified circumstances.
- Represent and explain phenomena with multiple types of models—for example, represent molecules with 3-D models or with bond diagrams—and move flexibly between model types when different ones are most useful for different purposes.
- Discuss the limitations and precision of a model as the representation of a system, process, or design and suggest ways in which the model might be improved to better fit available evidence or better reflect a design's specifications. Refine a model in light of empirical evidence or criticism to improve its quality and explanatory power.
- Use (provided) computer simulations or simulations developed with simple simulation tools as a tool for understanding and investigating aspects of a system, particularly those not readily visible to the naked eye.
- Make and use a model to test a design, or aspects of a design, and to compare the effectiveness of different design solutions.

PROGRESSION

Modeling can begin in the earliest grades, with students' models progressing from concrete "pictures" and/or physical scale models (e.g., a toy car) to more abstract representations of relevant relationships in later grades, such as a diagram representing forces on a particular object in a system. Students should be asked to use diagrams, maps, and other abstract models as tools that enable them to elaborate on their own ideas or findings and present them to others. Young students should be encouraged to devise pictorial and simple graphical representations of the findings of their investigations and to use these models in developing their explanations of what occurred.

More sophisticated types of models should increasingly be used across the grades, both in instruction and curriculum materials, as students progress through their science education. The quality of a student-developed model will be highly dependent on prior knowledge and skill and also on the student's understanding of the system being modeled, so students should be expected to refine their models as their understanding develops. Curricula will need to stress the role of models explicitly and provide students with modeling tools (e.g., Model-It, agent-based modeling such as NetLogo, spreadsheet models), so that students come to value this core practice and develop a level of facility in constructing and applying appropriate models.

Science and Engineering Practice 3: Planning and Carrying Out Investigations

Scientists and engineers investigate and observe the world with essentially two goals: (1) to systematically describe the world and (2) to develop and test theories and explanations of how the world works. In the first, careful observation and description often lead to identification of features that need to be explained or questions that need to be explored.

The second goal requires investigations to test explanatory models of the world and their predictions and whether the inferences suggested by these models are supported by data. Planning and designing such investigations require the ability to design experimental or observational inquiries that are appropriate to answering the question being asked or testing a hypothesis that has been formed. This process begins by identifying the relevant variables and considering how they might be observed, measured, and controlled (constrained by the experimental design to take particular values).

Planning for controls is an important part of the design of an investigation. In laboratory experiments, it is critical to decide which variables are to be treated as results or outputs and thus left to vary at will and which are to be treated as input conditions and hence controlled. In many cases, particularly in the case of field observations, such planning involves deciding what can be controlled and how to collect different samples of data under different conditions, even though not all conditions are under the direct control of the investigator.

Decisions must also be made about what measurements should be taken, the level of accuracy required, and the kinds of instrumentation best suited to making such measurements. As in other forms of inquiry, the key issue is one of precision—the goal is to measure the variable as accurately as possible and reduce sources of error. The investigator must therefore decide what constitutes a sufficient level of precision and what techniques can be used to reduce both random and systematic errors.

GOALS

By grade 12, students should be able to

- Formulate a question that can be investigated within the scope of the classroom, school laboratory, or field with available resources and, when appropriate, frame a hypothesis (that is, a possible explanation that predicts a particular and stable outcome) based on a model or theory.
- Decide what data are to be gathered, what tools are needed to do the gathering, and how measurements will be recorded.
- Decide how much data are needed to produce reliable measurements and consider any limitations on the precision of the data.
- Plan experimental or field-research procedures, identifying relevant independent and dependent variables and, when appropriate, the need for controls.

- Consider possible confounding variables or effects and ensure that the investigation's design has controlled for them.

PROGRESSION

Students need opportunities to design investigations so that they can learn the importance of such decisions as what to measure, what to keep constant, and how to select or construct data collection instruments that are appropriate to the needs of an inquiry. They also need experiences that help them recognize that the laboratory is not the sole domain for legitimate scientific inquiry and that, for many scientists (e.g., Earth scientists, ethologists, ecologists), the "laboratory" is the natural world where experiments are conducted and data are collected in the field.

In the elementary years, students' experiences should be structured to help them learn to define the features to be investigated, such as patterns that suggest causal relationships (e.g., What features of a ramp affect the speed of a given ball as it leaves the ramp?). The plan of the investigation, what trials to make, and how to record information about them then needs to be refined iteratively as students recognize from their experiences the limitations of their original plan. These investigations can be enriched and extended by linking them to engineering design projects—for example, how can students apply what they have learned about ramps to design a track that makes a ball travel a given distance, go around a loop, or stop on an uphill slope. From the earliest grades, students should have opportunities to carry out careful and systematic investigations, with appropriately supported prior experiences that develop their ability to observe and measure and to record data using appropriate tools and instruments.

Students should have opportunities to plan and carry out several different kinds of investigations during their K–12 years. At all levels, they should engage in investigations that range from those structured by the teacher—in order to expose an issue or question that they would be unlikely to explore on their own (e.g., measuring specific properties of materials)—to those that emerge from students' own questions. As they become more sophisticated, students also should have opportunities not only to identify questions to be researched but also to decide what data are to be gathered, what variables should be controlled, what tools or instruments are needed to gather and record data in an appropriate format, and eventually to consider how to incorporate measurement error in analyzing data.

Older students should be asked to develop a hypothesis that predicts a particular and stable outcome and to explain their reasoning and justify their choice. By high school, any hypothesis should be based on a well-developed model or theory. In addition, students should be able to recognize that it is not always possible to control variables and that other methods can be used in such cases—for example, looking for correlations (with the understanding that correlations do not necessarily imply causality).

Science and Engineering Practice 4: Analyzing and Interpreting Data

Once collected, data must be presented in a form that can reveal any patterns and relationships and that allows results to be communicated to others. Because raw data as such have little meaning, a major practice of scientists is to organize and interpret data through tabulating, graphing, or statistical analysis. Such analysis can bring out the meaning of data—and their relevance—so that they may be used as evidence.

Engineers, too, make decisions based on evidence that a given design will work; they rarely rely on trial and error. Engineers often analyze a design by creating a model or prototype and collecting extensive data on how it performs, including under extreme conditions. Analysis of this kind of data not only informs design decisions and enables the prediction or assessment of performance but also helps define or clarify problems, determine economic feasibility, evaluate alternatives, and investigate failures.

Spreadsheets and databases provide useful ways of organizing data, especially large data sets. The identification of relationships in data is aided by a range of tools, including tables, graphs, and mathematics. Tables permit major features of a large body of data to be summarized in a conveniently accessible form, graphs offer a means of visually summarizing data, and mathematics is essential for expressing relationships between different variables in the data set (see Practice 5 for further discussion of mathematics). Modern computer-based visualization tools often allow data to be displayed in varied forms and thus for learners to engage interactively with data in their analyses. In addition, standard statistical techniques can help to reduce the effect of error in relating one variable to another.

Students need opportunities to analyze large data sets and identify correlations. Increasingly, such data sets—involving temperature, pollution levels, and other scientific measurements—are available on the internet. Moreover, information technology enables the capture of data beyond the classroom at all hours of the day. Such data sets extend the range of students' experiences and help to illuminate this important practice of analyzing and interpreting data.

GOALS

By grade 12, students should be able to

- Analyze data systematically, either to look for salient patterns or to test whether data are consistent with an initial hypothesis.
- Recognize when data are in conflict with expectations and consider what revisions in the initial model are needed.
- Use spreadsheets, databases, tables, charts, graphs, statistics, mathematics, and information and computer technology to collate, summarize, and display data and to explore relationships between variables, especially those representing input and output.

- Evaluate the strength of a conclusion that can be inferred from any data set, using appropriate grade-level mathematical and statistical techniques.
- Recognize patterns in data that suggest relationships worth investigating further. Distinguish between causal and correlational relationships.
- Collect data from physical models and analyze the performance of a design under a range of conditions.

PROGRESSION

At the elementary level, students need support to recognize the need to record observations—whether in drawings, words, or numbers—and to share them with others. As they engage in scientific inquiry more deeply, they should begin to collect categorical or numerical data for presentation in forms that facilitate interpretation, such as tables and graphs. When feasible, computers and other digital tools should be introduced as a means of enabling this practice.

In middle school, students should have opportunities to learn standard techniques for displaying, analyzing, and interpreting data; such techniques include different types of graphs, the identification of outliers in the data set, and averaging to reduce the effects of measurement error. Students should also be asked to explain why these techniques are needed.

As students progress through various science classes in high school and their investigations become more complex, they need to develop skill in additional techniques for displaying and analyzing data, such as x-y scatterplots or cross-tabulations to express the relationship between two variables. Students should be helped to recognize that they may need to explore more than one way to display their data in order to identify and present significant features. They also need opportunities to use mathematics and statistics to analyze features of data such as covariation. Also at the high school level, students should have the opportunity to use a greater diversity of samples of scientific data and to use computers or other digital tools to support this kind of analysis.

Students should be expected to use some of these same techniques in engineering as well. When they do so, it is important that they are made cognizant of the purpose of the exercise—that any data they collect and analyze are intended to help validate or improve a design or decide on an optimal solution.

Science and Engineering Practice 5: Using Mathematics and Computational Thinking

Mathematics and computational tools are central to science and engineering. Mathematics enables the numerical representation of variables, the symbolic representation of relationships between physical entities, and the prediction of outcomes. Mathematics provides powerful models for describing and predicting such phenomena as atomic structure, gravitational forces, and quantum mechanics. Since the mid-20th century, computational theories, information and computer technologies, and algorithms have revolutionized virtually all scientific and engineering fields. These tools and strategies allow scientists and engineers to collect and analyze large data sets, search for distinctive patterns, and identify relationships and significant features in ways that were previously impossible. They also provide powerful new techniques for employing mathematics to model complex phenomena—for example, the circulation of carbon dioxide in the atmosphere and ocean.

Mathematics and computation can be powerful tools when brought to bear in a scientific investigation. Mathematics serves pragmatic functions as a tool—both a communicative function, as one of the languages of science, and a structural function, which allows for logical deduction. Mathematics enables ideas to be expressed in a precise form and enables the identification of new ideas about the physical world. For example, the concept of the equivalence of mass and energy emerged from the mathematical analysis conducted by Einstein, based on the premises of special relativity. The contemporary understanding of electromagnetic waves emerged from Maxwell's mathematical analysis of the behavior of electric and magnetic fields. Modern theoretical physics is so heavily imbued with mathematics that it would make no sense to try to divide it into mathematical and nonmathematical parts. In much of modern science, predictions and inferences have a probabilistic nature, so understanding the mathematics of probability and of statistically derived inferences is an important part of understanding science.

Computational tools enhance the power of mathematics by enabling calculations that cannot be carried out analytically. For example, they allow the development of simulations, which combine mathematical representations of multiple underlying phenomena to model the dynamics of a complex system. Computational methods are also potent tools for visually representing data, and they can show the results of calculations or simulations in ways that allow the exploration of patterns.

Engineering, too, involves mathematical and computational skills. For example, structural engineers create mathematical models of bridge and building designs, based on physical laws, to test their performance, probe their structural limits, and assess whether they can be completed within acceptable budgets. Virtually any engineering design raises issues that require computation for their resolution.

Although there are differences in how mathematics and computational thinking are applied in science and in engineering, mathematics often brings these two fields together by enabling engineers to apply the mathematical form of scientific theories and by enabling scientists to use powerful information technologies designed by engineers. Both kinds of professionals can thereby accomplish investigations and analyses and build complex models, which might otherwise be out of the question.

Mathematics (including statistics) and computational tools are essential for data analysis, especially for large data sets. The abilities to view data from different perspectives and with different graphical representations, to test relationships between variables, and to explore the interplay of diverse external conditions all require mathematical skills that are enhanced and extended with computational skills.

GOALS

By grade 12, students should be able to

- Recognize dimensional quantities and use appropriate units in scientific applications of mathematical formulas and graphs.
- Express relationships and quantities in appropriate mathematical or algorithmic forms for scientific modeling and investigations.
- Recognize that computer simulations are built on mathematical models that incorporate underlying assumptions about the phenomena or systems being studied.
- Use simple test cases of mathematical expressions, computer programs, or simulations—that is, compare their outcomes with what is known about the real world—to see if they "make sense."
- Use grade-level-appropriate understanding of mathematics and statistics in analyzing data.

PROGRESSION

Increasing students' familiarity with the role of mathematics in science is central to developing a deeper understanding of how science works. As soon as students learn to count, they can begin using numbers to find or describe patterns in nature. At appropriate grade levels, they should learn to use such instruments as rulers, protractors, and thermometers for the measurement of variables that are best represented by a continuous numerical scale, to apply mathematics to interpolate values, and to identify features—such as maximum, minimum, range, average, and median—of simple data sets.

A significant advance comes when relationships are expressed using equalities first in words and then in algebraic symbols—for example, shifting from distance traveled equals velocity multiplied by time elapsed to $s = vt$. Students should have opportunities to explore how such symbolic representations can be used to represent data, to predict outcomes, and eventually to derive further relationships using mathematics. Students should gain experience in using computers to record measurements taken with computer-connected probes or instruments, thereby recognizing how this process allows multiple measurements to be made rapidly and recurrently. Likewise, students should gain experience in using computer programs to transform their data between various tabular and graphical forms, thereby aiding in the identification of patterns.

Students should thus be encouraged to explore the use of computers for data analysis, using simple data sets, at an early age. For example, they could use spreadsheets to record data and then perform simple and recurring calculations from those data, such as the calculation of average speed from measurements of positions at multiple times. Later work should introduce them to the use of mathematical relationships to build simple computer models, using appropriate supporting programs or information and computer technology tools. As students progress in their understanding of mathematics and computation, at every level the science classroom should be a place where these tools are progressively exploited.

Science and Engineering Practice 6: Constructing Explanations and Designing Solutions

Because science seeks to enhance human understanding of the world, scientific theories are developed to provide explanations aimed at illuminating the nature of particular phenomena, predicting future events, or making inferences about past events. Science has developed explanatory theories, such as the germ theory of disease, the Big Bang theory of the origin of the universe, and Darwin's theory of the evolution of species. Although their role is often misunderstood—the informal use of the word "theory," after all, can mean a guess—*scientific* theories are constructs based on significant bodies of knowledge and evidence, are revised in light of new evidence, and must withstand significant scrutiny by the scientific community before they are widely accepted and applied. Theories are not mere guesses, and they are especially valued because they provide explanations for multiple instances.

In science, the term "hypothesis" is also used differently than it is in everyday language. A scientific hypothesis is neither a scientific theory nor a guess; it is a plausible explanation for an observed phenomenon that can predict what will happen in a given situation. A hypothesis is made based on existing theoretical understanding relevant to the situation and often also on a specific model for the system in question.

Scientific explanations are accounts that link scientific theory with specific observations or phenomena—for example, they explain observed relationships between variables and describe the mechanisms that support cause-and-effect inferences about them. Very often the theory is first represented by a specific model for the situation in question, and then a model-based explanation is developed. For example, if one understands the theory of how oxygen is obtained, transported, and utilized in the body, then a model of the circulatory system can be developed and used to explain why heart rate and breathing rate increase with exercise.

Engaging students with standard scientific explanations of the world—helping them to gain an understanding of the major ideas that science has developed—is a central aspect of science education. Asking students to demonstrate their own understanding of the implications of a scientific idea by developing their own explanations of phenomena, whether based on observations they have made or models they have developed, engages them in an essential part of the process by which conceptual change can occur. Explanations in science are a natural for such pedagogical uses, given their inherent appeals to simplicity, analogy, and empirical data (which may even be in the form of a thought experiment). And explanations are especially valuable for the classroom because of, rather than in spite of, the fact that there often are competing explanations offered for the same phenomenon—for example, the recent gradual rise in the mean surface temperature on Earth. Deciding on the best explanation is a matter of argument that is resolved by how well any given explanation fits with all available data, how much it simplifies what would seem to be complex, and whether it produces a sense of understanding.

Because scientists achieve their own understanding by building theories and theory-based explanations with the aid of models and representations and by drawing on data and evidence, students should also develop some facility in constructing model- or evidence-based explanations. This is an essential step in building their own understanding of phenomena, in gaining greater appreciation of the explanatory power of the scientific theories that they are learning about in class, and in acquiring greater insight into how scientists operate.

In engineering, the goal is a design rather than an explanation. The process of developing a design is iterative and systematic, as is the process of developing an explanation or a theory in science. Engineers' activities, however, have elements that are distinct from those of scientists. These elements include specifying constraints and criteria for desired qualities of the solution, developing a design plan, producing and testing models or prototypes, selecting among alternative design features to optimize the achievement of design criteria, and refining design ideas based on the performance of a prototype or simulation.

GOALS

By grade 12, students should be able to

- Construct their own explanations of phenomena using their knowledge of accepted scientific theory and linking it to models and evidence.
- Use primary or secondary scientific evidence and models to support or refute an explanatory account of a phenomenon.
- Offer causal explanations appropriate to their level of scientific knowledge.
- Identify gaps or weaknesses in explanatory accounts (their own or those of others).

In their experience of engineering, students should have the opportunity to

- Solve design problems by appropriately applying their scientific knowledge.
- Undertake design projects, engaging in all steps of the design cycle and producing a plan that meets specific design criteria.
- Construct a device or implement a design solution.
- Evaluate and critique competing design solutions based on jointly developed and agreed-on design criteria.

PROGRESSION FOR EXPLANATION

Early in their science education, students need opportunities to engage in constructing and critiquing explanations. They should be encouraged to develop explanations of what they observe when conducting their own investigations and to evaluate their own and others' explanations for consistency with the evidence. For example, observations of the owl pellets they dissect should lead them to produce an explanation of owls' eating habits based on inferences made from what they find.

As students' knowledge develops, they can begin to identify and isolate variables and incorporate the resulting observations into their explanations of phenomena. Using their measurements of how one factor does or does not affect another, they can develop causal accounts to explain what they observe. For example, in investigating the conditions under which plants grow fastest, they may notice that the plants die when kept in the dark and seek to develop an explanation for this finding. Although the explanation at this level may be as simple as "plants die in the dark because they need light in order to live and grow," it provides a basis for further questions and deeper understanding of how plants utilize light that can be developed in later grades. On the basis of comparison of their explanation with their observations, students can appreciate that an explanation such as "plants need light to grow" fails to explain why they die when no water is provided. They should be encouraged to revisit their initial ideas and produce more complete explanations that account for more of their observations.

By the middle grades, students recognize that many of the explanations of science rely on models or representations of entities that are too small to see or too large to visualize. For example, explaining why the temperature of water does not increase beyond 100°C when heated requires students to envisage water as consisting of microscopic particles and that the energy provided by heating can allow fast-moving particles to escape despite the force of attraction holding the particles together. In the later stages of their education, students should also progress to using mathematics or simulations to construct an explanation for a phenomenon.

PROGRESSION FOR DESIGN

In some ways, children are natural engineers. They spontaneously build sand castles, dollhouses, and hamster enclosures, and they use a variety of tools and materials for their own playful purposes. Thus a common elementary school activity is to challenge children to use tools and materials provided in class to solve a specific challenge, such as constructing a bridge from paper and tape and testing it until failure occurs. Children's capabilities to design structures can then be enhanced by having them pay attention to points of failure and asking them to create and test redesigns of the bridge so that it is stronger. Furthermore, design activities should not be limited just to structural engineering but should also include projects that reflect other areas of engineering, such as the need to design a traffic pattern for the school parking lot or a layout for planting a school garden box.

In middle school, it is especially beneficial to engage students in engineering design projects in which they are expected to apply what they have recently learned in science—for example, using their now-familiar concepts of ecology to solve problems related to a school garden. Middle school students should also have opportunities to plan and carry out full engineering design projects in which they define problems in terms of criteria and constraints, research the problem to deepen their relevant knowledge, generate and test possible solutions, and refine their solutions through redesign.

At the high school level, students can undertake more complex engineering design projects related to major local, national, or global issues. Increased emphasis should be placed on researching the nature of the given problems, on reviewing others' proposed solutions, on weighing the strengths and weaknesses of various alternatives, and on discerning possibly unanticipated effects.

Science and Engineering Practice 7: Engaging in Argument From Evidence

Whether they concern new theories, proposed explanations of phenomena, novel solutions to technological problems, or fresh interpretations of old data, scientists and engineers use reasoning and argumentation to make their case. In science, the production of knowledge is dependent on a process of reasoning that requires a scientist to make a justified claim about the world. In response, other scientists attempt to identify the claim's weaknesses and limitations. Their arguments can be based on deductions from premises, on inductive generalizations of existing patterns, or on inferences about the best possible explanation. Argumentation is also needed to resolve questions involving, for example, the best experimental design, the most appropriate techniques of data analysis, or the best interpretation of a given data set.

In short, science is replete with arguments that take place both informally, in lab meetings and symposia, and formally, in peer review. Historical case studies of the origin and development of a scientific idea show how a new idea is often difficult to accept and has to be argued for—archetypal examples are the Copernican idea that Earth travels around the Sun and Darwin's ideas about the origin of species. Over time, ideas that survive critical examination even in the light of new data attain consensual acceptance in the community, and by this process of discourse and argument science maintains its objectivity and progress.

The knowledge and ability to detect "bad science" are requirements both for the scientist and the citizen. Scientists must make critical judgments about their own work and that of their peers, and the scientist and the citizen alike must make evaluative judgments about the validity of science-related media reports and their implications for people's own lives and society. Becoming a critical consumer of science is fostered by opportunities to use critique and evaluation to judge the merits of any scientifically based argument.

In engineering, reasoning and argument are essential to finding the best possible solution to a problem. At an early design stage, competing ideas must be compared (and possibly combined) to achieve an initial design, and the choices are made through argumentation about the merits of the various ideas pertinent to the design goals. At a later stage in the design process, engineers test their potential solution, collect data, and modify their design in an iterative manner. The results of such efforts are often presented as evidence to argue about the strengths and weaknesses of a particular design. Although the forms of argumentation are similar, the criteria employed in engineering are often quite different from those of science. For example, engineers might use cost-benefit analysis, an analysis of risk, an appeal to aesthetics, or predictions about market reception to justify why one design is better than another—or why an entirely different course of action should be followed.

GOALS

By grade 12, students should be able to

- Construct a scientific argument showing how data support a claim.

- Identify possible weaknesses in scientific arguments, appropriate to the students' level of knowledge, and discuss them using reasoning and evidence.
- Identify flaws in their own arguments and modify and improve them in response to criticism.
- Recognize that the major features of scientific arguments are claims, data, and reasons and distinguish these elements in examples.
- Explain the nature of the controversy in the development of a given scientific idea, describe the debate that surrounded its inception, and indicate why one particular theory succeeded.
- Explain how claims to knowledge are judged by the scientific community today and articulate the merits and limitations of peer review and the need for independent replication of critical investigations.
- Read media reports of science or technology in a critical manner so as to identify their strengths and weaknesses.

PROGRESSION

The study of science and engineering should produce a sense of the process of argument necessary for advancing and defending a new idea or an explanation of a phenomenon and the norms for conducting such arguments. In that spirit, students should argue for the explanations they construct, defend their interpretations of the associated data, and advocate for the designs they propose. Meanwhile, they should learn how to evaluate critically the scientific arguments of others and present counterarguments. Learning to argue scientifically offers students not only an opportunity to use their scientific knowledge in justifying an explanation and in identifying the weaknesses in others' arguments but also to build their own knowledge and understanding. Constructing and critiquing arguments are both a core process of science and one that supports science education, as research suggests that interaction with others is the most cognitively effective way of learning.

Young students can begin by constructing an argument for their own interpretation of the phenomena they observe and of any data they collect. They need instructional support to go beyond simply making claims—that is, to include reasons or references to evidence and to begin to distinguish evidence from opinion. As they grow in their ability to construct scientific arguments, students can draw on a wider range of reasons or evidence, so that their arguments become more sophisticated. In addition, they should be expected to discern what aspects of the evidence are potentially significant for supporting or refuting a particular argument.

Students should begin learning to critique by asking questions about their own findings and those of others. Later, they should be expected to identify possible weaknesses in either data or an argument and explain why their criticism is justified. As they become more adept at arguing and critiquing, they should be introduced to the language needed to talk about argument, such as claim, reason, data, etc. Exploration of historical episodes in science can provide opportunities for students to identify the ideas, evidence, and arguments of professional scientists. In so doing, they should be encouraged to recognize the criteria used to judge claims for new knowledge and the formal means by which scientific ideas are evaluated today. In particular, they should see how the practice of peer review and independent verification of claimed experimental results help to maintain objectivity and trust in science.

Science and Engineering Practice 8: Obtaining, Evaluating, and Communicating Information

Being literate in science and engineering requires the ability to read and understand their literatures. Science and engineering are ways of knowing that are represented and communicated by words, diagrams, charts, graphs, images, symbols, and mathematics. Reading, interpreting, and producing text are fundamental practices of science in particular, and they constitute at least half of engineers' and scientists' total working time.

Even when students have developed grade-level-appropriate reading skills, reading in science is often challenging to students for three reasons. First, the jargon of science texts is essentially unfamiliar; together with their often extensive use of, for example, the passive voice and complex sentence structure, many find these texts inaccessible. Second, science texts must be read so as to extract information accurately. Because the precise meaning of each word or clause may be important, such texts require a mode of reading that is quite different from reading a novel or even a newspaper. Third, science texts are multimodal, using a mix of words, diagrams, charts, symbols, and mathematics to communicate. Thus understanding science texts requires much more than simply knowing the meanings of technical terms.

Communicating in written or spoken form is another fundamental practice of science; it requires scientists to describe observations precisely, clarify their thinking, and justify their arguments. Because writing is one of the primary means of communicating in the scientific community, learning how to produce scientific texts is as essential to developing an understanding of science as learning how to draw is to appreciating the skill of the visual artist. Indeed, the new *Common Core State Standards for English Language Arts & Literacy in History/Social Studies, Science, and Technical Subjects* recognize that reading and writing skills are essential to science; the formal inclusion in this framework of this science practice reinforces and expands on that view. Science simply cannot advance if scientists are unable to communicate their findings clearly and persuasively. Communication occurs in a variety of formal venues, including peer-reviewed journals, books, conference presentations, and carefully constructed websites; it occurs as well through informal means, such as discussions, e-mail messages, phone calls, and blogs. New technologies have extended communicative practices, enabling multidisciplinary collaborations across the globe that place even more emphasis on reading and writing. Increasingly, too, scientists are required to engage in dialogues with lay audiences about their work, which requires especially good communication skills.

Being a critical consumer of science and the products of engineering, whether as a lay citizen or a practicing scientist or an engineer, also requires the ability to read or view reports about science in the press or on the internet and to recognize the salient science, identify sources of error and methodological flaws, and distinguish observations from inferences, arguments from explanations, and claims from evidence. All of these are constructs learned from engaging in a critical discourse around texts. Engineering proceeds in a similar manner because engineers need to communicate ideas and find and exchange information—for example, about new techniques or new uses of existing tools and materials. As in science, engineering communication involves not just written and spoken language; many engineering ideas are best communicated through sketches, diagrams, graphs, models, and products. Also in wide use are handbooks, specific to particular engineering fields, that provide detailed information, often in tabular form, on how best to formulate design solutions to commonly encountered engineering tasks. Knowing how to seek and use such informational resources is an important part of the engineer's skill set.

GOALS

By grade 12, students should be able to

- Use words, tables, diagrams, and graphs (whether in hard copy or electronically), as well as mathematical expressions, to communicate their understanding or to ask questions about a system under study.
- Read scientific and engineering text, including tables, diagrams, and graphs, commensurate with their scientific knowledge and explain the key ideas being communicated.
- Recognize the major features of scientific and engineering writing and speaking and be able to produce written and illustrated text or oral presentations that communicate their own ideas and accomplishments.
- Engage in a critical reading of primary scientific literature (adapted for classroom use) or of media reports of science and discuss the validity and reliability of the data, hypotheses, and conclusions.

PROGRESSION

Any education in science and engineering needs to develop students' ability to read and produce domain-specific text. As such, every science or engineering lesson is in part a language lesson, particularly reading and producing the genres of texts that are intrinsic to science and engineering.

Students need sustained practice and support to develop the ability to extract the meaning of scientific text from books, media reports, and other forms of scientific communication because the form of this text is initially unfamiliar—expository rather than narrative, often linguistically dense, and reliant on precise logical flows. Students should be able to interpret meaning from text, to produce text in which written language and diagrams are used to express scientific ideas, and to engage in extended discussion about those ideas.

From the very start of their science education, students should be asked to engage in the communication of science, especially regarding the investigations they are conducting and the observations they are making. Careful description of observations and clear statement of ideas, with the ability to both refine a statement in response to questions and to ask questions of others to achieve clarification of what is being said begin at the earliest grades. Beginning in upper elementary and middle school, the ability to interpret written materials becomes more important. Early work on reading science texts should also include explicit instruction and practice in interpreting tables, diagrams, and charts and coordinating information conveyed by them with information in written text. Throughout their science education, students are continually introduced to new terms, and the meanings of those terms can be learned only through opportunities to use and apply them in their specific contexts. Not only must students learn technical terms but also more general academic language, such as "analyze" or "correlation," which are not part of most students' everyday vocabulary and thus need specific elaboration if they are to make sense of scientific text. It follows that to master the reading of scientific material, students need opportunities to engage with such text and to identify its major features; they cannot be expected simply to apply reading skills learned elsewhere to master this unfamiliar genre effectively.

Students should write accounts of their work, using journals to record observations, thoughts, ideas, and models. They should be encouraged to create diagrams and to represent data and observations with plots and tables, as well as with written text, in these journals. They should also begin to produce reports or posters that present their work to others. As students begin to read and write more texts, the particular genres of scientific text—a report of an investigation, an explanation with supporting argumentation, an experimental procedure—will need to be introduced and their purpose explored. Furthermore, students should have opportunities to engage in discussion about observations and explanations and to make oral presentations of their results and conclusions as well as to engage in appropriate discourse with other students by asking questions and discussing issues raised in such presentations. Because the spoken language of such discussions and presentations is as far from their everyday language as scientific text is from a novel, the development both of written and spoken scientific explanation/ argumentation needs to proceed in parallel.

In high school, these practices should be further developed by providing students with more complex texts and a wider range of text materials, such as technical reports or scientific literature on the internet. Moreover, students need opportunities to read and discuss general media reports with a critical eye and to read appropriate samples of adapted primary literature to begin seeing how science is communicated by science practitioners.

In engineering, students likewise need opportunities to communicate ideas using appropriate combinations of sketches, models, and language. They should also create drawings to test concepts and communicate detailed plans; explain and critique models of various sorts, including scale models and prototypes; and present the results of simulations, not only regarding the planning and development stages but also to make compelling presentations of their ultimate solutions.

Crosscutting Concepts*

Crosscutting Concept 1: Patterns

Patterns exist everywhere—in regularly occurring shapes or structures and in repeating events and relationships. For example, patterns are discernible in the symmetry of flowers and snowflakes, the cycling of the seasons, and the repeated base pairs of DNA. Noticing patterns is often a first step to organizing and asking scientific questions about why and how the patterns occur.

One major use of pattern recognition is in classification, which depends on careful observation of similarities and differences; objects can be classified into groups on the basis of similarities of visible or microscopic features or on the basis of similarities of function. Such classification is useful in codifying relationships and organizing a multitude of objects or processes into a limited number of groups. Patterns of similarity and difference and the resulting classifications may change, depending on the scale at which a phenomenon is being observed. For example, isotopes of a given element are different—they contain different numbers of neutrons—but from the perspective of chemistry they can be classified as equivalent because they have identical patterns of chemical interaction. Once patterns and variations have been noted, they lead to questions; scientists seek explanations for observed patterns and for the similarity and diversity within them. Engineers often look for and analyze patterns, too. For example, they may diagnose patterns of failure of a designed system under test in order to improve the design, or they may analyze patterns of daily and seasonal use of power to design a system that can meet the fluctuating needs.

The ways in which data are represented can facilitate pattern recognition and lead to the development of a mathematical representation, which can then be used as a tool in seeking an underlying explanation for what causes the pattern to occur. For example, biologists studying changes in population abundance of several different species in an ecosystem can notice the correlations between increases and decreases for different species by plotting all of them on the same graph and can eventually find a mathematical expression of the interdependences and food-web relationships that cause these patterns.

PROGRESSION

Human beings are good at recognizing patterns; indeed, young children begin to recognize patterns in their own lives well before coming to school. They observe, for example, that the Sun and the Moon follow different patterns of appearance in the sky. Once they are students,

---

* Excerpted with permission from: National Research Council (NRC). 2012. *A framework for K–12 science education: Practices, crosscutting concepts, and core ideas.* Washington, DC: National Academies Press.

it is important for them to develop ways to recognize, classify, and record patterns in the phenomena they observe. For example, elementary students can describe and predict the patterns in the seasons of the year; they can observe and record patterns in the similarities and differences between parents and their offspring. Similarly, they can investigate the characteristics that allow classification of animal types (e.g., mammals, fish, insects), of plants (e.g., trees, shrubs, grasses), or of materials (e.g., wood, rock, metal, plastic).

These classifications will become more detailed and closer to scientific classifications in the upper elementary grades, when students should also begin to analyze patterns in rates of change—for example, the growth rates of plants under different conditions. By middle school, students can begin to relate patterns to the nature of microscopic and atomic-level structure—for example, they may note that chemical molecules contain particular ratios of different atoms. By high school, students should recognize that different patterns may be observed at each of the scales at which a system is studied. Thus classifications used at one scale may fail or need revision when information from smaller or larger scales is introduced (e.g., classifications based on DNA comparisons vs. those based on visible characteristics).

Crosscutting Concept 2: Cause and Effect: Mechanism and Prediction

Many of the most compelling and productive questions in science are about why or how something happens. Any tentative answer, or "hypothesis," that A causes B requires a model for the chain of interactions that connect A and B. For example, the notion that diseases can be transmitted by a person's touch was initially treated with skepticism by the medical profession for lack of a plausible mechanism. Today infectious diseases are well understood as being transmitted by the passing of microscopic organisms (bacteria or viruses) between an infected person and another. A major activity of science is to uncover such causal connections, often with the hope that understanding the mechanisms will enable predictions and, in the case of infectious diseases, the design of preventive measures, treatments, and cures.

Repeating patterns in nature, or events that occur together with regularity, are clues that scientists can use to start exploring causal, or cause-and-effect, relationships, which pervade all the disciplines of science and at all scales. For example, researchers investigate cause-and-effect mechanisms in the motion of a single object, specific chemical reactions, population changes in an ecosystem or a society, and the development of holes in the polar ozone layers. Any application of science, or any engineered solution to a problem, is dependent on understanding the cause-and-effect relationships between events; the quality of the application or solution often can be improved as knowledge of the relevant relationships is improved.

Identifying cause and effect may seem straightforward in simple cases, such as a bat hitting a ball, but in complex systems causation can be difficult to tease out. It may be conditional, so that A can cause B only if some other factors are in place or within a certain numerical range. For example, seeds germinate and produce plants but only when the soil is sufficiently moist and warm. Frequently, causation can be described only in a probabilistic fashion—that is, there is some likelihood that one event will lead to another, but a specific outcome cannot be guaranteed. For example, one can predict the fraction of a collection of identical atoms that will undergo radioactive decay in a certain period but not the exact time at which a given atom decays.

One assumption of all science and engineering is that there is a limited and universal set of fundamental physical interactions that underlie all known forces and hence are a root part of any causal chain, whether in natural or designed systems. Such "universality" means that the physical laws underlying all processes are the same everywhere and at all times; they depend on gravity, electromagnetism, or weak and strong nuclear interactions. Underlying all biological processes—the inner workings of a cell or even of a brain—are particular physical and chemical processes. At the larger scale of biological systems, the universality of life manifests itself in a common genetic code.

Causation invoked to explain larger scale systems must be consistent with the implications of what is known about smaller scale processes within the system, even though new features may emerge at large scales that cannot be predicted from knowledge of smaller scales. For example, although knowledge of atoms is not sufficient to predict the genetic code, the replication of genes must be understood as a molecular-level process. Indeed, the ability to model causal processes in complex multipart systems arises from this fact; modern computational codes incorporate relevant smaller scale relationships into the model of the larger system, integrating multiple factors in a way that goes well beyond the capacity of the human brain.

In engineering, the goal is to design a system to cause a desired effect, so cause-and-effect relationships are as much a part of engineering as of science. Indeed, the process of design is a good place to help students begin to think in terms of cause and effect, because they must understand the underlying causal relationships in order to devise and explain a design that can achieve a specified objective.

One goal of instruction about cause and effect is to encourage students to see events in the world as having understandable causes, even when these causes are beyond human control. The ability to distinguish between scientific causal claims and nonscientific causal claims is also an important goal.

PROGRESSION

In the earliest grades, as students begin to look for and analyze patterns—whether in their observations of the world or in the relationships between different quantities in data (e.g., the sizes of plants over time)—they can also begin to consider what might be causing these patterns and relationships and design tests that gather more evidence to support or refute their ideas. By the upper elementary grades, students should have developed the habit of routinely asking about cause-and effect relationships in the systems they are studying, particularly when something occurs that is, for them, unexpected. The questions "How did that happen?" or "Why did that happen?" should move toward "What mechanisms caused that to happen?" and "What conditions were critical for that to happen?"

In middle and high school, argumentation starting from students' own explanations of cause and effect can help them appreciate standard scientific theories that explain the causal mechanisms in the systems under study. Strategies for this type of instruction include asking students to argue from evidence when attributing an observed phenomenon to a specific cause. For example, students exploring why the population of a given species is shrinking will look for evidence in the ecosystem of factors that lead to food shortages, overpredation, or other factors in the habitat related to survival; they will provide an argument for how these and other observed changes affect the species of interest.

Crosscutting Concept 3: Scale, Proportion, and Quantity

In thinking scientifically about systems and processes, it is essential to recognize that they vary in size (e.g., cells, whales, galaxies), in time span (e.g., nanoseconds, hours, millennia), in the amount of energy flowing through them (e.g., lightbulbs, power grids, the Sun), and in the relationships between the scales of these different quantities. The understanding of relative magnitude is only a starting point. As noted in *Benchmarks for Science Literacy*, "The large idea is that the way in which things work may change with scale. Different aspects of nature change at different rates with changes in scale, and so the relationships among them change, too." Appropriate understanding of scale relationships is critical as well to engineering—no structure could be conceived, much less constructed, without the engineer's precise sense of scale.

From a human perspective, one can separate three major scales at which to study science: (1) macroscopic scales that are directly observable—that is, what one can see, touch, feel, or manipulate; (2) scales that are too small or fast to observe directly; and (3) those that are too large or too slow. Objects at the atomic scale, for example, may be described with simple models, but the size of atoms and the number of atoms in a system involve magnitudes that are difficult to imagine. At the other extreme, science deals in scales that are equally difficult to imagine because they are so large—continents that move, for example, and galaxies in which the nearest star is 4 years away traveling at the speed of light. As size scales change, so do time scales. Thus, when considering large entities such as mountain ranges, one typically needs to consider change that occurs over long periods. Conversely, changes in a small-scale system, such as a cell, are viewed over much shorter times. However, it is important to recognize that processes that occur locally and on short time scales can have long-term and large-scale impacts as well.

In forming a concept of the very small and the very large, whether in space or time, it is important to have a sense not only of relative scale sizes but also of what concepts are meaningful at what scale. For example, the concept of solid matter is meaningless at the subatomic scale, and the concept that light takes time to travel a given distance becomes more important as one considers large distances across the universe.

Understanding scale requires some insight into measurement and an ability to think in terms of orders of magnitude—for example, to comprehend the difference between one in a hundred and a few parts per billion. At a basic level, in order to identify something as bigger or smaller than something else—and how much bigger or smaller—a student must appreciate the units used to measure it and develop a feel for quantity.

The ideas of ratio and proportionality as used in science can extend and challenge students' mathematical understanding of these concepts. To appreciate the relative magnitude of some properties or processes, it may be necessary to grasp the relationships among different types of quantities—for example, speed as the ratio of distance traveled to time taken, density as a ratio of mass to volume. This use of ratio is quite different than a ratio of numbers describing fractions of a pie. Recognition of such relationships among different quantities is a key step in forming mathematical models that interpret scientific data.

PROGRESSION

The concept of scale builds from the early grades as an essential element of understanding phenomena. Young children can begin understanding scale with objects, space, and time related to their world and with explicit scale models and maps. They may discuss relative scales—the biggest and smallest, hottest and coolest, fastest and slowest—without reference to particular units of measurement.

Typically, units of measurement are first introduced in the context of length, in which students can recognize the need for a common unit of measure—even develop their own before being introduced to standard units—through appropriately constructed experiences. Engineering design activities involving scale diagrams and models can support students in developing facility with this important concept.

Once students become familiar with measurements of length, they can expand their understanding of scale and of the need for units that express quantities of weight, time, temperature, and other variables. They can also develop an understanding of estimation across scales and contexts, which is important for making sense of data. As students become more sophisticated, the use of estimation can help them not only to develop a sense of the size and time scales relevant to various objects, systems, and processes but also to consider whether a numerical result sounds reasonable. Students acquire the ability as well to move back and forth between models at various scales, depending on the question being considered. They should develop a sense of the powers-of-10 scales and what phenomena correspond to what scale, from the size of the nucleus of an atom to the size of the galaxy and beyond.

Well-designed instruction is needed if students are to assign meaning to the types of ratios and proportional relationships they encounter in science. Thus the ability to recognize mathematical relationships between quantities should begin developing in the early grades with students' representations of counting (e.g., leaves on a branch), comparisons of amounts (e.g., of flowers on different plants), measurements (e.g., the height of a plant), and the ordering of quantities such as number, length, and weight. Students can then explore more sophisticated mathematical representations, such as the use of graphs to represent data collected. The interpretation of these graphs may be, for example, that a plant gets bigger as time passes or that the hours of daylight decrease and increase across the months.

As students deepen their understanding of algebraic thinking, they should be able to apply it to examine their scientific data to predict the effect of a change in one variable on another, for example, or to appreciate the difference between linear growth and exponential growth. As their thinking advances, so too should their ability to recognize and apply more complex mathematical and statistical relationships in science. A sense of numerical quantity is an important part of the general "numeracy" (mathematics literacy) that is needed to interpret such relationships.

Crosscutting Concept 4: Systems and System Models

As noted in the *National Science Education Standards*, "The natural and designed world is complex; it is too large and complicated to investigate and comprehend all at once. Scientists and students learn to define small portions for the convenience of investigation. The units of investigations can be referred to as 'systems.' A system is an organized group of related objects or components that form a whole. Systems can consist, for example, of organisms, machines, fundamental particles, galaxies, ideas, and numbers. Systems have boundaries, components, resources, flow, and feedback."

Although any real system smaller than the entire universe interacts with and is dependent on other (external) systems, it is often useful to conceptually isolate a single system for study. To do this, scientists and engineers imagine an artificial boundary between the system in question and everything else. They then examine the system in detail while treating the effects of things outside the boundary as either forces acting on the system or flows of matter and energy across it—for example, the gravitational force due to Earth on a book lying on a table or the carbon dioxide expelled by an organism. Consideration of flows into and out of the system is a crucial element of system design. In the laboratory or even in field research, the extent to which a system under study can be physically isolated or external conditions controlled is an important element of the design of an investigation and interpretation of results.

Often, the parts of a system are interdependent, and each one depends on or supports the functioning of the system's other parts. Yet the properties and behavior of the whole system can be very different from those of any of its parts, and large systems may have emergent properties, such as the shape of a tree, that cannot be predicted in detail from knowledge about the components and their interactions. Things viewed as subsystems at one scale may themselves be viewed as whole systems at a smaller scale. For example, the circulatory system can be seen as an entity in itself or as a subsystem of the entire human body; a molecule can be studied as a stable configuration of atoms but also as a subsystem of a cell or a gas.

An explicit model of a system under study can be a useful tool not only for gaining understanding of the system but also for conveying it to others. Models of a system can range in complexity from lists and simple sketches to detailed computer simulations or functioning prototypes.

Models can be valuable in predicting a system's behaviors or in diagnosing problems or failures in its functioning, regardless of what type of system is being examined. A good system model for use in developing scientific explanations or engineering designs must specify not only the parts, or subsystems, of the system but also how they interact with one another. It must also specify the boundary of the system being modeled, delineating what is included in the model and what is to be treated as external. In a simple mechanical system, interactions among the parts are describable in terms of forces among them that cause changes in motion or physical stresses. In more complex systems, it is not always possible or useful to consider interactions at this detailed mechanical level, yet it is equally important to ask what interactions are occurring (e.g., predator-prey relationships in an ecosystem) and to recognize that they all involve transfers of energy, matter, and (in some cases) information among parts of the system.

Any model of a system incorporates assumptions and approximations; the key is to be aware of what they are and how they affect the model's reliability and precision. Predictions may be reliable but not precise or, worse, precise but not reliable; the degree of reliability and precision needed depends on the use to which the model will be put.

PROGRESSION

As science instruction progresses, so too should students' ability to analyze and model more complex systems and to use a broader variety of representations to explicate what they model. Their thinking about systems in terms of component parts and their interactions, as well as in terms of inputs, outputs, and processes, gives students a way to organize their knowledge of a system, to generate questions that can lead to enhanced understanding, to test aspects of their model of the system, and, eventually, to refine their model.

Starting in the earliest grades, students should be asked to express their thinking with drawings or diagrams and with written or oral descriptions. They should describe objects or organisms in terms of their parts and the roles those parts play in the functioning of the object or organism, and they should note relationships between the parts. Students should also be asked to create plans—for example, to draw or write a set of instructions for building something—that another child can follow. Such experiences help them develop the concept of a model of a system and realize the importance of representing one's ideas so that others can understand and use them.

As students progress, their models should move beyond simple renderings or maps and begin to incorporate and make explicit the invisible features of a system, such as interactions, energy flows, or matter transfers. Mathematical ideas, such as ratios and simple graphs, should be seen as tools for making more definitive models; eventually, students' models should incorporate a range of mathematical relationships among variables (at a level appropriate for grade-level mathematics) and some analysis of the patterns of those relationships. By high school, students should also be able to identify the assumptions and approximations that have been built into a model and discuss how they limit the precision and reliability of its predictions.

Instruction should also include discussion of the interactions *within* a system. As understanding deepens, students can move from a vague notion of interaction as one thing affecting another to more explicit realizations of a system's physical, chemical, biological, and social interactions and of their relative importance for the question at hand. Students' ideas about the interactions in a system and the explication of such interactions in their models should become more sophisticated in parallel with their understanding of the microscopic world (atoms, molecules, biological cells, microbes) and with their ability to interpret and use more complex mathematical relationships.

Modeling is also a tool that students can use in gauging their own knowledge and clarifying their questions about a system. Student-developed models may reveal problems or progress in their conceptions of the system, just as scientists' models do. Teaching students to explicitly craft and present their models in diagrams, words, and, eventually, in mathematical relationships serves three purposes. It supports them in clarifying their ideas and explanations and in considering any inherent contradictions; it allows other students the opportunity to critique and suggest revisions for the model; and it offers the teacher insights into those aspects of each student's understanding that are well founded and those that could benefit from further instructional attention. Likewise in engineering projects, developing systems thinking and system models supports critical steps in developing, sharing, testing, and refining design ideas.

Crosscutting Concept 5: Energy and Matter: Flows, Cycles, and Conservation

One of the great achievements of science is the recognition that, in any system, certain conserved quantities can change only through transfers into or out of the system. Such laws of conservation provide limits on what can occur in a system, whether human built or natural. This section focuses on two such quantities, matter and energy, whose conservation has important implications for the disciplines of science in this framework. The supply of energy and of each needed chemical element restricts a system's operation—for example, without inputs of energy (sunlight) and matter (carbon dioxide and water), a plant cannot grow. Hence, it is very informative to track the transfers of matter and energy within, into, or out of any system under study.

In many systems there also are cycles of various types. In some cases, the most readily observable cycling may be of matter—for example, water going back and forth between Earth's atmosphere and its surface and subsurface reservoirs. Any such cycle of matter also involves associated energy transfers at each stage, so to fully understand the water cycle, one must model not only how water moves between parts of the system but also the energy transfer mechanisms that are critical for that motion.

Consideration of energy and matter inputs, outputs, and flows or transfers within a system or process are equally important for engineering. A major goal in design is to maximize certain types of energy output while minimizing others, in order to minimize the energy inputs needed to achieve a desired task.

The ability to examine, characterize, and model the transfers and cycles of matter and energy is a tool that students can use across virtually all areas of science and engineering. And studying the *interactions* between matter and energy supports students in developing increasingly sophisticated conceptions of their role in any system. However, for this development to occur, there needs to be a common use of language about energy and matter across the disciplines in science instruction.

PROGRESSION

The core ideas of matter and energy and their development across the grade bands are spelled out in detail in Chapter 5 [of *A Framework for K–12 Science Education*, NRC 2012]. What is added in this crosscutting discussion is recognition that an understanding of these core ideas can be informative in examining systems in life science, earth and space science, and engineering contexts. Young children are likely to have difficulty studying the concept of energy in depth—everyday language surrounding energy contains many shortcuts that lead to misunderstandings. For this reason, the concept is not developed at all in K–2 and only very generally in grades 3–5. Instead, the elementary grades focus on recognition of conservation of matter and of the flow of matter into, out of, and within systems under study. The role of energy transfers in conjunction with these flows is not introduced until the middle grades and only fully developed by high school.

Clearly, incorrect beliefs—such as the perception that food or fuel is a form of energy—would lead to elementary grade students' misunderstanding of the nature of energy. Hence, although the necessity for food or fuel can be discussed, the language of energy needs to be used with care so as not to further establish such misconceptions. By middle school, a more precise idea of energy—for example, the understanding that food or fuel undergoes a chemical reaction with oxygen that releases stored energy—can emerge. The common misconceptions can be addressed with targeted instructional interventions (including student-led investigations), and appropriate terminology can be used in discussing energy across the disciplines.

Matter transfers are less fraught in this respect, but the idea of atoms is not introduced with any specificity until middle school. Thus, at the level of grades 3–5, matter flows and cycles can be tracked only in terms of the weight of the substances before and after a process occurs, such as sugar dissolving in water. Mass/weight distinctions and the idea of atoms and their conservation (except in nuclear processes) are taught in grades 6–8, with nuclear substructure and the related conservation laws for nuclear processes introduced in grades 9–12.

Crosscutting Concept 6: Structure and Function

As expressed by the National Research Council in 1996 and reiterated by the College Board in 2009, "Form and function are complementary aspects of objects, organisms, and systems in the natural and designed world. ... Understanding of form and function applies to different levels of organization. Function can be explained in terms of form and form can be explained in terms of function."

The functioning of natural and built systems alike depends on the shapes and relationships of certain key parts as well as on the properties of the materials from which they are made. A sense of scale is necessary in order to know what properties and what aspects of shape or material are relevant at a particular magnitude or in investigating particular phenomena—that is, the selection of an appropriate scale depends on the question being asked. For example, the substructures of molecules are not particularly important in understanding the phenomenon of pressure, but they are relevant to understanding why the ratio between temperature and pressure at constant volume is different for different substances.

Similarly, understanding how a bicycle works is best addressed by examining the structures and their functions at the scale of, say, the frame, wheels, and pedals. However, building a lighter bicycle may require knowledge of the properties (such as rigidity and hardness) of the materials needed for specific parts of the bicycle. In that way, the builder can seek less dense materials with appropriate properties; this pursuit may lead in turn to an examination of the atomic-scale structure of candidate materials. As a result, new parts with the desired properties, possibly made of new materials, can be designed and fabricated.

PROGRESSION

Exploration of the relationship between structure and function can begin in the early grades through investigations of accessible and visible systems in the natural and human-built world. For example, children explore how shape and stability are related for a variety of structures (e.g., a bridge's diagonal brace) or purposes (e.g., different animals get their food using different parts of their bodies). As children move through the elementary grades, they progress to understanding the relationships of structure and mechanical function (e.g., wheels and axles, gears). For upper-elementary students, the concept of matter having a substructure at a scale too small to see is related to properties of materials; for example, a model of a gas as a collection of moving particles (not further defined) may be related to observed properties of gases. Upper-elementary students can also examine more complex structures, such as subsystems of the human body, and consider the relationship of the shapes of the parts to their functions. By the middle grades, students begin to visualize, model, and apply their understanding of structure and function to more complex or less easily observable systems and processes (e.g., the structure of water and salt molecules and solubility, Earth's plate tectonics). For students in the middle grades, the concept of matter having a submicroscopic structure is related to properties of materials; for example, a model based on atoms and/or molecules and their motions may be used to explain the properties of solids, liquids, and gases or the evaporation and condensation of water.

As students develop their understanding of the relationships between structure and function, they should begin to apply this knowledge when investigating phenomena that are unfamiliar to them. They recognize that often the first step in deciphering how a system works is to examine in detail what it is made of and the shapes of its parts. In building something—say, a mechanical system—they likewise apply relationships of structure and function as critical elements of successful designs.

Crosscutting Concept 7: Stability and Change

"Much of science and mathematics has to do with understanding how change occurs in nature and in social and technological systems, and much of technology has to do with creating and controlling change," according to the American Association for the Advancement of Science. "Constancy, often in the midst of change, is also the subject of intense study in science."

Stability denotes a condition in which some aspects of a system are unchanging, at least at the scale of observation. Stability means that a small disturbance will fade away—that is, the system will stay in, or return to, the stable condition. Such stability can take different forms, with the simplest being a static equilibrium, such as a ladder leaning on a wall. By contrast, a system with steady inflows and outflows (i.e., constant conditions) is said to be in dynamic equilibrium. For example, a dam may be at a constant level with steady quantities of water coming in and out. Increase the inflow, and a new equilibrium level will eventually be reached if the outflow increases as well. At extreme flows, other factors may cause *dis*equilibrium; for example, at a low-enough inflow, evaporation may cause the level of the water to continually drop. Likewise, a fluid at a constant temperature can be in a steady state with constant chemical composition even though chemical reactions that change the composition in two opposite directions are occurring within it; change the temperature and it will reach a new steady state with a different composition.

A repeating pattern of cyclic change—such as the Moon orbiting Earth—can also be seen as a stable situation, even though it is clearly not static. Such a system has constant aspects, however, such as the distance from Earth to the Moon, the period of its orbit, and the pattern of phases seen over time.

In designing systems for stable operation, the mechanisms of external controls and internal "feedback" loops are important design elements; feedback is important to understanding natural systems as well. A feedback loop is any mechanism in which a condition triggers some action that causes a change in that same condition, such as the temperature of a room triggering the thermostatic control that turns the room's heater on or off. Feedback can stabilize a system (negative feedback—a thermostat in a cooling room triggers heating, but only until a particular temperature range is reached) or destabilize a system (positive feedback—a fire releases heat, which triggers the burning of more fuel, which causes the fire to continue to grow).

A system can be stable on a small time scale, but on a larger time scale it may be seen to be changing. For example, when looking at a living organism over the course of an hour or a day, it may maintain stability; over longer periods, the organism grows, ages, and eventually dies. For the development of larger systems, such as the variety of living species inhabiting Earth or the formation of a galaxy, the relevant time scales may be very long indeed; such processes occur over millions or even billions of years.

When studying a system's patterns of change over time, it is also important to examine what is unchanging. Understanding the feedback mechanisms that regulate the system's stability or that drive its instability provides insight into how the system may operate under various conditions. These mechanisms are important to evaluate when comparing different design options that address a particular problem.

Any system has a range of conditions under which it can operate in a stable fashion, as well as conditions under which it cannot function. For example, a particular living organism can survive only within a certain range of temperatures, and outside that span it will die. Thus elucidating what range of conditions can lead to a system's stable operation and what changes would destabilize it (and in what ways) is an important goal.

Note that stability is always a balance of competing effects; a small change in conditions or in a single component of the system can lead to runaway changes in the system if compensatory mechanisms are absent. Nevertheless, students typically begin with an idea of equilibrium as a static situation, and they interpret a lack of change in the system as an indication that nothing is happening. Thus they need guidance to begin to appreciate that stability can be the result of multiple opposing forces; they should be taught to identify the invisible forces—to appreciate the dynamic equilibrium—in a seemingly static situation, even one as simple as a book lying on a table.

An understanding of dynamic equilibrium is crucial to understanding the major issues in any complex system—for example, population dynamics in an ecosystem or the relationship between the level of atmospheric carbon dioxide and Earth's average temperature. Dynamic equilibrium is an equally important concept for understanding the physical forces in matter. Stable matter is a system of atoms in dynamic equilibrium.

For example, the stability of the book lying on the table depends on the fact that minute distortions of the table caused by the book's downward push on the table in turn cause changes in the positions of the table's atoms. These changes then alter the forces between those atoms, which lead to changes in the upward force on the book exerted by the table. The book continues to distort the table until the table's upward force on the book exactly balances the downward pull of gravity on the book. Place a heavy enough item on the table, however, and stability is not possible; the distortions of matter within the table continue to the macroscopic scale, and it collapses under the weight. Such seemingly simple, explicit, and visible examples of how change in some factor produces changes in the system can help to establish a mental model of dynamic equilibrium useful for thinking about more complex systems.

Understanding long-term changes—for example, the evolution of the diversity of species, the surface of Earth, or the structure of the universe—requires a sense of the requisite time scales for such changes to develop. Long time scales can be difficult for students to grasp, however. Part of their understanding should grow from an appreciation of how scientists investigate the nature of these processes—through the interplay of evidence and system modeling. Student-developed models that use comparative time scales can also be helpful; for example, if the history of Earth is scaled to 1 year (instead of the absolute measures in eons), students gain a more intuitive understanding of the relative durations of periods in the planet's evolution.

PROGRESSION

Even very young children begin to explore stability (as they build objects with blocks or climb on a wall) and change (as they note their own growth or that of a plant). The role of instruction in the early grades is to help students to develop some language for these concepts and apply it appropriately across multiple examples, so that they can ask such questions as "What could I change to make this balance better?" or "How fast did the plants grow?" One of the goals of discussion of stability and change in the elementary grades should be the recognition that it can be as important to ask why something does not change as why it does.

Likewise, students should come to recognize that both the regularities of a pattern over time and its variability are issues for which explanations can be sought. Examining these questions in different contexts (e.g., a model ecosystem such as a terrarium, the local weather, a design for a bridge) broadens students' understanding that stability and change are related and that a good model for a system must be able to offer explanations for both.

In middle school, as student's understanding of matter progresses to the atomic scale, so too should their models and their explanations of stability and change. Furthermore, they can begin to appreciate more subtle or conditional situations and the need for feedback to maintain stability. At the high school level, students can model more complex systems and comprehend more subtle issues of stability or of sudden or gradual change over time. Students at this level should also recognize that much of science deals with constructing historical explanations of how things evolved to be the way they are today, which involves modeling rates of change and conditions under which the system is stable or changes gradually, as well as explanations of any sudden change.

A Look at the *Next Generation Science Standards*

The *Next Generation Science Standards* (*NGSS*) differ from prior science standards in that they integrate three dimensions (science and engineering practices, disciplinary core ideas, and crosscutting concepts) into a single performance expectation and have intentional connections between performance expectations. The system architecture of *NGSS* highlights the performance expectations as well as each of the three integral dimensions and connections to other grade bands and subjects. The architecture involves a table with three main sections.

What Is Assessed (Performance Expectations)	**Foundation Box**	**Connection Box**
A performance expectation describes what students should be able to do at the end of instruction and incorporates a science and engineering practice, a disciplinary core idea (DCI), and a crosscutting concept from the foundation box. Performance expectations are not instructional strategies or objectives for a lesson. Instead, they are intended to guide the development of assessments. Groupings of performance expectations do not imply a preferred ordering for instruction—nor should all performance expectations under one topic necessarily be taught in one course. This section also contains *Clarification Statements* and *Assessment Boundary* statements that are meant to render additional support and clarity to the performance expectations.	The foundation box contains the learning goals that students should achieve. It is critical that science educators consider the foundation box an essential component when reading the *NGSS* and developing curricula. There are three main parts of the foundation box: science and engineering practices, DCIs, and crosscutting concepts, all of which are derived from *A Framework for K–12 Science Education*. During instruction, teachers will need to have students use multiple practices to help students understand the core ideas. Most topical groupings of performance expectations emphasize only a few practices or crosscutting concepts; however, all are emphasized within a grade band. The foundation box also contains learning goals for *Connections to Engineering, Technology, and Applications of Science* and *Connections to Nature of Science.*	The connection box identifies other topics in *NGSS* and in the *Common Core State Standards* (*CCSS*) that are relevant to the performance expectations in this topic. The connections to other DCIs in this grade level contains the codes for topics in other science disciplines that have corresponding disciplinary core ideas at the same grade level. The *Articulation of Disciplinary Core Ideas (DCIs)* across grade levels contains the codes of other science topics that either provide a foundation for student understanding of the core ideas in this standard (usually standards at prior grade levels) or build on the foundation provided by the core ideas in this standard (usually standards at subsequent grade levels). The *Connections to the Common Core State Standards* contains the descriptions of *CCSS*, in ELA/literacy and mathematics that align to the performance expectations.

Inside the *NGSS* Box

What Is Assessed

A collection of several performance expectations describing what students should be able to do at the end of instruction

Foundation Box

The practices, disciplinary core ideas, and crosscutting concepts from *A Framework for K–12 Science Education* that were used to form the performance expectations

Connection Box

Places elsewhere in *NGSS* or in the *Common Core State Standards* that have connections to the performance expectations on this page

Title

The title for a set of performance expectations is not necessarily unique and may be reused at several different grade levels.

MS-LS2 Ecosystems: Interactions, Energy, and Dynamics

Students who demonstrate understanding can:

MS-LS2-3. **Develop a model to describe the cycling of matter and flow of energy among living and nonliving parts of an ecosystem.**
[Clarification Statement: Emphasis is on describing the conservation of matter and flow of energy into and out of various ecosystems, and on defining the boundaries of the system.]
[Assessment Boundary: Assessment does not include the use of chemical reactions to describe the processes.]

MS-LS2-4. **Construct an argument supported by empirical evidence that changes to physical or biological components of an ecosystem affect populations.**
[Clarification Statement: Emphasis is on recognizing patterns in data and making warranted inferences about changes in populations, and on evaluating empirical evidence supporting arguments about changes to ecosystems.]

MS-LS2-5. **Evaluate competing design solutions for maintaining biodiversity and ecosystem services.***
[Clarification Statement: Examples of ecosystem services could include water purification, nutrient recycling, and prevention of soil erosion. Examples of design solution constraints could include scientific, economic, and social considerations.]

The performance expectations above were developed using the following elements from the NRC document *A Framework for K-12 Science Education*:

Science and Engineering Practices

Developing and Using Models
Modeling in 6–8 builds on K–5 experiences and progresses to developing, using, and revising models to describe, test, and predict more abstract phenomena and design systems.
- Develop a model to describe phenomena. (MS-LS2-3)

Engaging in Argument from Evidence
Engaging in argument from evidence in 6–8 builds on K–5 experiences and progresses to constructing a convincing argument that supports or refutes claims for either explanations or solutions about the natural and designed world(s).
- Construct an oral and written argument supported by empirical evidence and scientific reasoning to support or refute an explanation or a model for a phenomenon or a solution to a problem. (MS-LS2-4)
- Evaluate competing design solutions based on jointly developed and agreed-upon design criteria. (MS-LS2-5)

Connections to Nature of Science
Scientific Knowledge is Based on Empirical Evidence
- Science disciplines share common rules of obtaining and evaluating empirical evidence. (MS-LS2-4)

Disciplinary Core Ideas

LS2.B: Cycle of Matter and Energy Transfer in Ecosystems
- Food webs are models that demonstrate how matter and energy is transferred between producers, consumers, and decomposers as the three groups interact within an ecosystem. Transfers of matter into and out of the physical environment occur at every level. Decomposers recycle nutrients from dead plant or animal matter back to the soil in terrestrial environments or to the water in aquatic environments. The atoms that make up the organisms in an ecosystem are cycled repeatedly between the living and nonliving parts of the ecosystem. (MS-LS2-3)

LS2.C: Ecosystem Dynamics, Functioning, and Resilience
- Ecosystems are dynamic in nature; their characteristics can vary over time. Disruptions to any physical or biological component of an ecosystem can lead to shifts in all its populations. (MS-LS2-4)
- Biodiversity describes the variety of species found in Earth's terrestrial and oceanic ecosystems. The completeness or integrity of an ecosystem's biodiversity is often used as a measure of its health. (MS-LS2-5)

LS4.D: Biodiversity and Humans
- Changes in biodiversity can influence humans' resources, such as food, energy, and medicines, as well as ecosystem services that humans rely on—for example, water purification and recycling. (secondary to MS-LS2-5)

ETS1.B: Developing Possible Solutions
- There are systematic processes for evaluating solutions with respect to how well they meet the criteria and constraints of a problem. (secondary to MS-LS2-5)

Crosscutting Concepts

Energy and Matter
- The transfer of energy can be tracked as energy flows through a natural system. (MS-LS2-3)

Stability and Change
- Small changes in one part of a system might cause large changes in another part. (MS-LS2-4), (MS-LS2-5)

Connections to Engineering, Technology, and Applications of Science
Influence of Science, Engineering, and Technology on Society and the Natural World
- The use of technologies and any limitations on their use are driven by individual or societal needs, desires, and values; by the findings of scientific research; and by differences in such factors as climate, natural resources, and economic conditions. Thus technology use varies from region to region and over time. (MS-LS2-5)

Connections to Nature of Science
Scientific Knowledge Assumes an Order and Consistency in Natural Systems
- Science assumes that objects and events in natural systems occur in consistent patterns that are understandable through measurement and observation. (MS-LS2-3)

Science Addresses Questions About the Natural and Material World
- Scientific knowledge can describe the consequences of actions but does not necessarily prescribe the decisions that society takes. (MS-LS2-5)

Connections to other DCIs in this grade-band: **MS.PS1.B** (MS-LS2-3); **MS.LS4.C** (MS-LS2-4); **MS.LS4.D** (MS-LS2-4); **MS.ESS2.A** (MS-LS2-3),(MS-LS2-4); **MS.ESS3.C** (MS-LS2-4),(MS-LS2-5)

Articulation across grade-bands: **3.LS2.C** (MS-LS2-4); **3.LS4.D** (MS-LS2-4); **5.LS2.A** (MS-LS2-3); **5.LS2.B** (MS-LS2-3); **HS.PS3.B** (MS-LS2-3); **HS.LS1.C** (MS-LS2-3); **HS.LS2.A** (MS-LS2-5); **HS.LS2.B** (MS-LS2-3); **HS.LS2.C** (MS-LS2-4),(MS-LS2-5); **HS.LS4.C** (MS-LS2-4); **HS.LS4.D** (MS-LS2-4),(MS-LS2-5); **HS.ESS2.A** (MS-LS2-3); **HS.ESS2.E** (MS-LS2-4); **HS.ESS3.A** (MS-LS2-5); **HS.ESS3.B** (MS-LS2-4); **HS.ESS3.C** (MS-LS2-4),(MS-LS2-5); **HS.ESS3.D** (MS-LS2-5)

Common Core State Standards Connections:

ELA/Literacy –

RST.6-8.1	Cite specific textual evidence to support analysis of science and technical texts. (MS-LS2-4)
RST.6-8.8	Distinguish among facts, reasoned judgment based on research findings, and speculation in a text. (MS-LS2-5)
RI.8.8	Trace and evaluate the argument and specific claims in a text, assessing whether the reasoning is sound and the evidence is relevant and sufficient to support the claims. (MS-LS-4),(MS-LS2-5)
WHST.6-8.1	Write arguments to support claims with clear reasons and relevant evidence. (MS-LS2-4)
WHST.6-8.9	Draw evidence from literary or informational texts to support analysis, reflection, and research. (MS-LS2-4)
SL.8.5	Integrate multimedia and visual displays into presentations to clarify information, strengthen claims and evidence, and add interest. (MS-LS2-3)

Mathematics –

MP.4	Model with mathematics. (MS-LS2-5)
6.RP.A.3	Use ratio and rate reasoning to solve real-world and mathematical problems. (MS-LS2-5)
6.EE.C.9	Use variables to represent two quantities in a real-world problem that change in relationship to one another; write an equation to express one quantity, thought of as the dependent variable, in terms of the other quantity, thought of as the independent variable. Analyze the relationship between the dependent and independent variables using graphs and tables, and relate these to the equation. (MS-LS2-3)

Performance Expectations

A statement that combines science and engineering practices, disciplinary core ideas, and crosscutting concepts to describe how students can show what they have learned

Clarification Statement

A statement that supplies examples or additional clarification to the performance expectation

Assessment Boundary

A statement that provides guidance about the scope of the performance expectation at a particular grade level.

Engineering Connection (*)

An asterisk indicates a performance expectation that integrates traditional science content with engineering through a practice or core idea.

Science and Engineering Practices

Activities that scientists and engineers engage in to either understand the world or solve a problem

Disciplinary Core Ideas

Concepts in science and engineering that have broad importance within and across disciplines as well as relevance in people's lives

Crosscutting Concepts

Ideas, such as Patterns and Cause and Effect, that are not specific to any one discipline but cut across them all

Connections to Engineering, Technology, and Applications of Science

These connections are drawn from the disciplinary core ideas for engineering, technology, and applications of science in the *Framework*.

Connections to Nature of Science

Connections are listed in either the practices or the crosscutting concepts section of the foundation box.

Codes for Performance Expectations

Every performance expectation has a unique code, and items in the foundation box and connection box reference this code. In the connections to *Common Core*, italics indicate a potential connection rather than a required connection.

NGSS Organized by Topic

Level		Life Science	Earth and Space Science	Physical Science	Engineering
Elementary School	K	K. Interdependent Relationships in Ecosystems: Animals, Plants, and Their Environment	K. Weather and Climate	K. Forces and Interactions: Pushes and Pulls	K–2. Engineering Design
	1	1. Structure, Function, and Information Processing	1. Space Systems: Patterns and Cycles	1. Waves: Light and Sound	
	2	2. Interdependent Relationships in Ecosystems	2. Earth's Systems: Processes That Shape the Earth	2. Structure and Properties of Matter	
	3	3. Interdependent Relationships in Ecosystems 3. Inheritance and Variation of Traits: Life Cycles and Traits	3. Weather and Climate	3. Forces and Interactions	3–5. Engineering Design
	4	4. Structure, Function, and Information Processing	4. Earth's Systems: Processes that Shape the Earth	4. Energy 4. Waves: Waves and Information	
	5	5. Matter and Energy in Organisms and Ecosystems	5. Earth's Systems 5. Space Systems: Stars and the Solar System	5. Structure and Properties of Matter	
Middle School		MS. Structure, Function, and Information Processing MS. Matter and Energy in Organisms and Ecosystems MS. Interdependent Relationships in Ecosystems MS. Natural Selection and Adaptations MS. Growth, Development, and Reproduction of Organisms	MS. Space Systems MS. History of Earth MS. Earth's Systems MS. Weather and Climate MS. Human Impacts	MS. Structure and Properties of Matter MS. Chemical Reactions MS. Forces and Interactions MS. Energy MS. Waves and Electromagnetic Radiation	MS. Engineering Design
High School		HS. Structure and Function HS. Inheritance and Variation of Traits HS. Matter and Energy in Organisms and Ecosystems HS. Interdependent Relationships in Ecosystems HS. Natural Selection and Evolution	HS. Space Systems HS. History of Earth HS. Earth's Systems HS. Weather and Climate HS. Human Sustainability	HS. Structure and Properties of Matter HS. Chemical Reactions HS. Forces and Interactions HS. Energy HS. Waves and Electromagnetic Radiation	HS. Engineering Design

NGSS Organized by Disciplinary Core Ideas

Level		Life Science	Earth and Space Science	Physical Science	Engineering
Elementary School	K	K-LS-1 From Molecules to Organisms: Structures and Processes	K-ESS-2 Earth's Systems K-ESS-3 Earth and Human Activity	K-PS-2 Motion and Stability: Forces and Interactions K-PS-3 Energy	K–2-ETS-1 Engineering Design
	1	1-LS-1 From Molecules to Organisms: Structures and Processes 1-LS-3 Heredity: Inheritance and Variation of Traits	1-ESS-1 Earth's Place in the Universe	1-PS-4 Waves and Their Applications in Technologies for Information Transfer	
	2	2-LS-2 Ecosystems: Interactions, Energy, and Dynamics 2-LS-4 Biological Evolution: Unity and Diversity	2-ESS-1 Earth's Place in the Universe 2-ESS-2 Earth's Systems	2-PS-1 Matter and Its Interactions	
	3	3-LS-1 From Molecules to Organisms: Structures and Processes 3-LS-2 Ecosystems: Interactions, Energy, and Dynamics 3-LS-3 Heredity: Inheritance and Variation of Traits 3-LS-4 Biological Evolution: Unity and Diversity	3-ESS-2 Earth's Systems 3-ESS-3 Earth and Human Activity	3-PS-2 Motion and Stability: Forces and Interactions	3–5-ETS-1 Engineering Design
	4	4-LS-1 From Molecules to Organisms: Structures and Processes	4-ESS-1 Earth's Place in the Universe 4-ESS-2 Earth's Systems 4-ESS-3 Earth and Human Activity	4-PS-3 Energy 4-PS-4 Waves and Their Applications in Technologies for Information Transfer	
	5	5-LS-1 From Molecules to Organisms: Structures and Processes 5-LS-2 Ecosystems: Interactions, Energy, and Dynamics	5-ESS-1 Earth's Place in the Universe 5-ESS-2 Earth's Systems 5-ESS-3 Earth and Human Activity	5-PS-1 Matter and Its Interactions 5-PS-2 Motion and Stability: Forces and Interactions 5-PS-3 Energy	
Middle School		MS-LS-1 From Molecules to Organisms: Structures and Processes MS-LS-2 Ecosystems: Interactions, Energy, and Dynamics MS-LS-3 Heredity: Inheritance and Variation of Traits MS-LS-4 Biological Evolution: Unity and Diversity	MS-ESS-1 Earth's Place in the Universe MS-ESS-2 Earth's Systems MS-ESS-3 Earth and Human Activity	MS-PS-1 Matter and Its Interactions MS-PS-2 Motion and Stability: Forces and Interactions MS-PS-3 Energy MS-PS-4 Waves and Their Applications in Technologies for Information Transfer	MS-ETS-1 Engineering Design
High School		HS-LS-1 From Molecules to Organisms: Structures and Processes HS-LS-2 Ecosystems: Interactions, Energy, and Dynamics HS-LS-3 Heredity: Inheritance and Variation of Traits HS-LS-4 Biological Evolution: Unity and Diversity	HS-ESS-1 Earth's Place in the Universe HS-ESS-2 Earth's Systems HS-ESS-3 Earth and Human Activity	HS-PS-1 Matter and Its Interactions HS-PS-2 Motion and Stability: Forces and Interactions HS-PS-3 Energy HS-PS-4 Waves and Their Applications in Technologies for Information Transfer	HS-ETS-1 Engineering Design

Commonalities Among the Practices in Science, Mathematics, and English Language Arts (ELA)

Source: Based on work by Tina Cheuk, *http://ell.stanford.edu/teaching_resources*

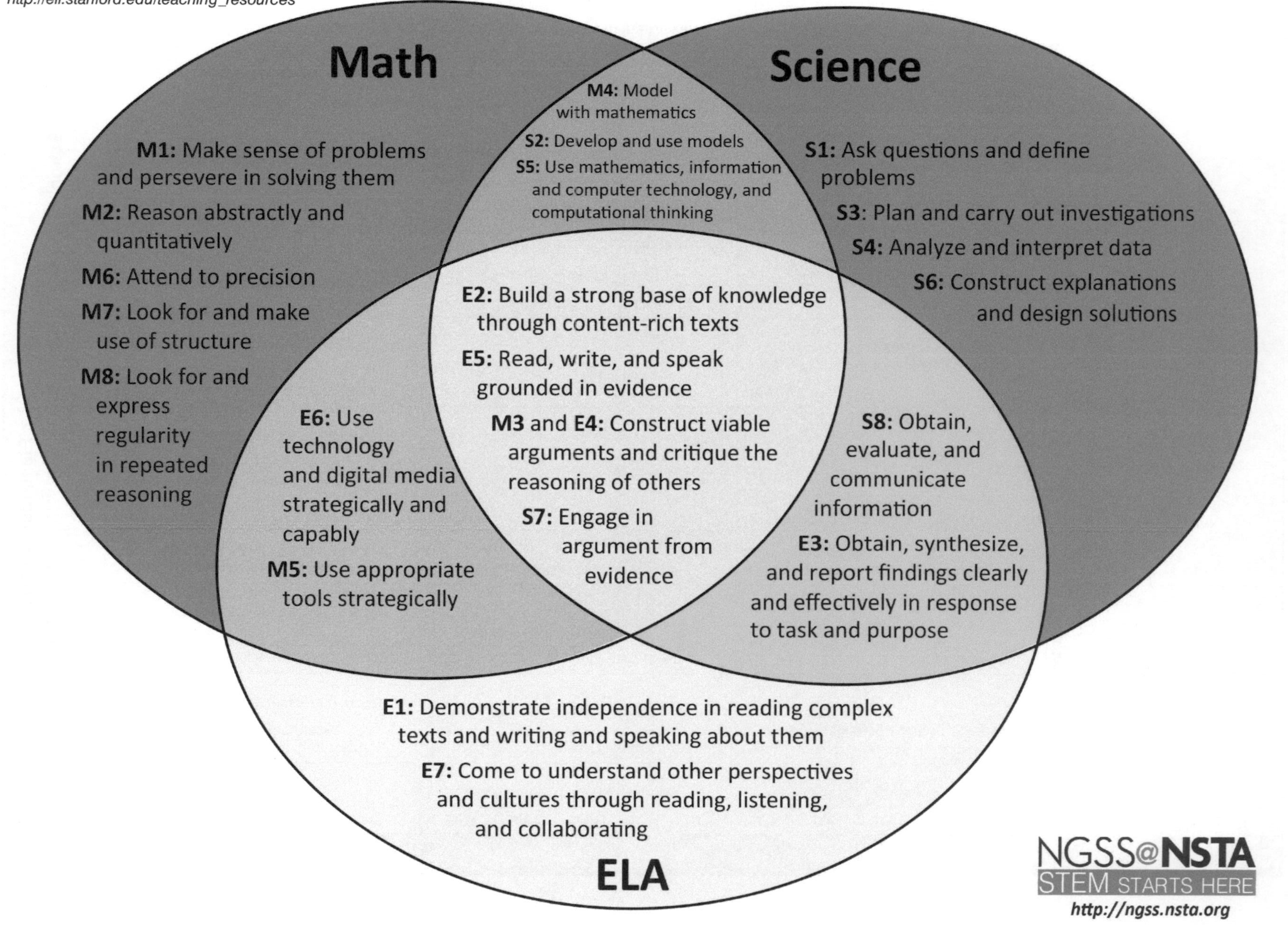

Practices in Mathematics, Science, and English Language Arts*

Mathematics	Science	English Language Arts
M1. Make sense of problems and persevere in solving them.	**S1.** Asking questions (for science) and defining problems (for engineering).	**E1.** They demonstrate independence.
M2. Reason abstractly and quantitatively.	**S2.** Developing and using models.	**E2.** They build strong content knowledge.
M3. Construct viable arguments and critique the reasoning of others.	**S3.** Planning and carrying out investigations.	**E3.** They respond to the varying demands of audience, task, purpose, and discipline.
M4. Model with mathematics.	**S4.** Analyzing and interpreting data.	**E4.** They comprehend as well as critique.
M5. Use appropriate tools strategically.	**S5.** Using mathematics and computational thinking.	**E5.** They value evidence.
M6. Attend to precision.	**S6.** Constructing explanations (for science) and designing solutions (for engineering).	**E6.** They use technology and digital media strategically and capably.
M7. Look for and make use of structure.	**S7.** Engaging in argument from evidence.	**E7.** They come to understand other perspectives and cultures.
M8. Look for and express regularity in repeated reasoning.	**S8.** Obtaining, evaluating, and communicating information.	

* The *Common Core State Standards, English Language Arts* uses the term *student capacities* rather than the term *practices* used in *Common Core State Standards, Mathematics* and the *Next Generation Science Standards*.

CHAPTER 2

K–12 Progressions

Science and Engineering Practices: Asking Questions and Defining Problems

A practice of science is to ask and refine questions that lead to descriptions and explanations of how the natural and designed world works and which can be empirically tested. Engineering questions clarify problems to determine criteria for successful solutions and identify constraints to solve problems about the designed world. Both scientists and engineers also ask questions to clarify ideas.

K–2 Condensed Practices	3–5 Condensed Practices	6–8 Condensed Practices	9–12 Condensed Practices
Asking questions and defining problems in K–2 builds on prior experiences and progresses to simple descriptive questions that can be tested.	Asking questions and defining problems in 3–5 builds on K–2 experiences and progresses to specifying qualitative relationships.	Asking questions and defining problems in 6–8 builds on K–5 experiences and progresses to specifying relationships between variables, clarifying arguments and making models.	Asking questions and defining problems in 9–12 builds on K–8 experiences and progresses to formulating, refining, and evaluating empirically testable questions and design problems using models and simulations.
• Ask questions based on observations to find more information about the natural and/or designed world(s).	• Ask questions about what would happen if a variable is changed.	• Ask questions that arise from careful observation of phenomena, models, or unexpected results, to clarify and/or seek additional information. • Ask questions to identify and/or clarify evidence and/or the premise(s) of an argument. • Ask questions to determine relationships between independent and dependent variables and relationships in models. • Ask questions to clarify and/or refine a model, an explanation, or an engineering problem.	• Ask questions that arise from careful observation of phenomena, or unexpected results, to clarify and/or seek additional information. • Ask questions that arise from examining models or a theory, to clarify and/or seek additional information and relationships. • Ask questions to determine relationships, including quantitative relationships, between independent and dependent variables. • Ask questions to clarify and refine a model, an explanation, or an engineering problem.
• Ask and/or identify questions that can be answered by an investigation.	• Identify scientific (testable) and non-scientific (non-testable) questions. • Ask questions that can be investigated and predict reasonable outcomes based on patterns such as cause-and-effect relationships.	• Ask questions that require sufficient and appropriate empirical evidence to answer. • Ask questions that can be investigated within the scope of the classroom, outdoor environment, and museums and other public facilities with available resources and, when appropriate, frame a hypothesis based on observations and scientific principles.	• Evaluate a question to determine if it is testable and relevant. • Ask questions that can be investigated within the scope of the school laboratory, research facilities, or field (e.g., outdoor environment) with available resources and, when appropriate, frame a hypothesis based on a model or theory.
• N/A	• N/A	• Ask questions that challenge the premise(s) of an argument or the interpretation of a data set.	• Ask and/or evaluate questions that challenge the premise(s) of an argument, the interpretation of a data set, or the suitability of the design.
• Define a simple problem that can be solved through the development of a new or improved object or tool.	• Use prior knowledge to describe problems that can be solved. • Define a simple design problem that can be solved through the development of an object, tool, process, or system and includes several criteria for success and constraints on materials, time, or cost.	• Define a design problem that can be solved through the development of an object, tool, process, or system and includes multiple criteria and constraints, including scientific knowledge that may limit possible solutions.	• Define a design problem that involves the development of a process or system with interacting components and criteria and constraints that may include social, technical, and/or environmental considerations.

N/A = Not applicable for this grade range

Science and Engineering Practices: Developing and Using Models

A practice of both science and engineering is to use and construct models as helpful tools for representing ideas and explanations. These tools include diagrams, drawings, physical replicas, mathematical representations, analogies, and computer simulations. Modeling tools are used to develop questions, predictions, and explanations; analyze and identify flaws in systems; and communicate ideas. Models are used to build and revise scientific explanations and proposed engineered systems. Measurements and observations are used to revise models and designs.

K–2 Condensed Practices	3–5 Condensed Practices	6–8 Condensed Practices	9–12 Condensed Practices
Modeling in K–2 builds on prior experiences and progresses to include using and developing models (e.g., diagram, drawing, physical replica, diorama, dramatization, or storyboard) that represent concrete events or design solutions.	Modeling in 3–5 builds on K–2 experiences and progresses to building and revising simple models and using models to represent events and design solutions.	Modeling in 6–8 builds on K–5 experiences and progresses to developing, using, and revising models to describe, test, and predict more abstract phenomena and design systems.	Modeling in 9–12 builds on K–8 experiences and progresses to using, synthesizing, and developing models to predict and show relationships among variables between systems and their components in the natural and designed world(s).
• Distinguish between a model and the actual object, process, and/or events the model represents. • Compare models to identify common features and differences.	• Identify limitations of models.	• Evaluate limitations of a model for a proposed object or tool.	• Evaluate merits and limitations of two different models of the same proposed tool, process, mechanism, or system in order to select or revise a model that best fits the evidence or design criteria. • Design a test of a model to ascertain its reliability.
• Develop and/or use a model to represent amounts, relationships, relative scales (bigger, smaller), and/or patterns in the natural and designed world(s).	• Collaboratively develop and/or revise a model based on evidence that shows the relationships among variables for frequent and regular occurring events. • Develop a model using an analogy, example, or abstract representation to describe a scientific principle or design solution. • Develop and/or use models to describe and/or predict phenomena.	• Develop or modify a model—based on evidence—to match what happens if a variable or component of a system is changed. • Use and/or develop a model of simple systems with uncertain and less predictable factors. • Develop and/or revise a model to show the relationships among variables, including those that are not observable but predict observable phenomena. • Develop and/or use a model to predict and/or describe phenomena. • Develop a model to describe unobservable mechanisms.	• Develop, revise, and/or use a model based on evidence to illustrate and/or predict the relationships between systems or between components of a system. • Develop and/or use multiple types of models to provide mechanistic accounts and/or predict phenomena, and move flexibly between model types based on merits and limitations.
• Develop a simple model based on evidence to represent a proposed object or tool.	• Develop a diagram or simple physical prototype to convey a proposed object, tool, or process. • Use a model to test cause-and-effect relationships or interactions concerning the functioning of a natural or designed system.	• Develop and/or use a model to generate data to test ideas about phenomena in natural or designed systems, including those representing inputs and outputs, and those at unobservable scales.	• Develop a complex model that allows for manipulation and testing of a proposed process or system. • Develop and/or use a model (including mathematical and computational) to generate data to support explanations, predict phenomena, analyze systems, and/or solve problems.

Science and Engineering Practices: Planning and Carrying Out Investigations

Scientists and engineers plan and carry out investigations in the field or laboratory, working collaboratively as well as individually. Their investigations are systematic and require clarifying what counts as data and identifying variables or parameters. Engineering investigations identify the effectiveness, efficiency, and durability of designs under different conditions.

K–2 Condensed Practices	3–5 Condensed Practices	6–8 Condensed Practices	9–12 Condensed Practices
Planning and carrying out investigations to answer questions or test solutions to problems in K–2 builds on prior experiences and progresses to simple investigations, based on fair tests, which provide data to support explanations or design solutions.	Planning and carrying out investigations to answer questions or test solutions to problems in 3–5 builds on K–2 experiences and progresses to include investigations that control variables and provide evidence to support explanations or design solutions.	Planning and carrying out investigations in 6–8 builds on K–5 experiences and progresses to include investigations that use multiple variables and provide evidence to support explanations or solutions.	Planning and carrying out investigations in 9–12 builds on K–8 experiences and progresses to include investigations that provide evidence for and test conceptual, mathematical, physical, and empirical models.
• With guidance, plan and conduct an investigation in collaboration with peers (for K). • Plan and conduct an investigation collaboratively to produce data to serve as the basis for evidence to answer a question.	• Plan and conduct an investigation collaboratively to produce data to serve as the basis for evidence, using fair tests in which variables are controlled and the number of trials considered.	• Plan an investigation individually and collaboratively, and in the design identify independent and dependent variables and controls, what tools are needed to do the gathering, how measurements will be recorded, and how many data are needed to support a claim. • Conduct an investigation and/or evaluate and/or revise the experimental design to produce data to serve as the basis for evidence that meet the goals of the investigation.	• Plan an investigation or test a design individually and collaboratively to produce data to serve as the basis for evidence as part of building and revising models, supporting explanations for phenomena, or testing solutions to problems. Consider possible variables or effects and evaluate the confounding investigation's design to ensure variables are controlled. • Plan and conduct an investigation individually and collaboratively to produce data to serve as the basis for evidence, and in the design decide on types, how much, and accuracy of data needed to produce reliable measurements and consider limitations on the precision of the data (e.g., number of trials, cost, risk, time); refine the design accordingly. • Plan and conduct an investigation or test a design solution in a safe and ethical manner including considerations of environmental, social, and personal impacts.
• Evaluate different ways of observing and/or measuring a phenomenon to determine which way can answer a question.	• Evaluate appropriate methods and/or tools for collecting data.	• Evaluate the accuracy of various methods for collecting data.	• Select appropriate tools to collect, record, analyze, and evaluate data.
• Make observations (firsthand or from media) and/or measurements to collect data that can be used to make comparisons. • Make observations (firsthand or from media) and/or measurements of a proposed object or tool or solution to determine if it solves a problem or meets a goal. • Make predictions based on prior experiences.	• Make observations and/or measurements to produce data to serve as the basis for evidence for an explanation of a phenomenon or test a design solution. • Make predictions about what would happen if a variable changes. • Test two different models of the same proposed object, tool, or process to determine which better meets criteria for success.	• Collect and produce data to serve as the basis for evidence to answer scientific questions or test design solutions under a range of conditions. • Collect data about the performance of a proposed object, tool, process, or system under a range of conditions.	• Make directional hypotheses that specify what happens to a dependent variable when an independent variable is manipulated. • Manipulate variables and collect data about a complex model of a proposed process or system to identify failure points or improve performance relative to criteria for success or other variables.

Science and Engineering Practices: Analyzing and Interpreting Data

Scientific investigations produce data that must be analyzed to derive meaning. Because data patterns and trends are not always obvious, scientists use a range of tools—including tabulation, graphical interpretation, visualization, and statistical analysis—to identify the significant features and patterns in the data. Scientists identify sources of error in the investigations and calculate the degree of certainty in the results. Modern technology makes the collection of large data sets much easier, providing secondary sources for analysis. Engineering investigations include analysis of data collected in the tests of designs. This allows comparison of different solutions and determines how well each meets specific design criteria—that is, which design best solves the problem within given constraints. Like scientists, engineers require a range of tools to identify patterns within data and interpret the results. Advances in science make analysis of proposed solutions more efficient and effective.

K–2 Condensed Practices	3–5 Condensed Practices	6–8 Condensed Practices	9–12 Condensed Practices
Analyzing data in K–2 builds on prior experiences and progresses to collecting, recording, and sharing observations.	Analyzing data in 3–5 builds on K–2 experiences and progresses to introducing quantitative approaches to collecting data and conducting multiple trials of qualitative observations. When possible and feasible, digital tools should be used.	Analyzing data in 6–8 builds on K–5 experiences and progresses to extending quantitative analysis to investigations, distinguishing between correlation and causation, and basic statistical techniques of data and error analysis.	Analyzing data in 9–12 builds on K–8 experiences and progresses to introducing more detailed statistical analysis, the comparison of data sets for consistency, and the use of models to generate and analyze data.
• Record information (observations, thoughts, and ideas). • Use observations (firsthand or from media) to describe patterns and/or relationships in the natural and designed world in order to answer scientific questions and solve problems. • Compare predictions (based on prior experiences) to what occurred (observable events).	• Represent data in tables and/or various graphical displays (bar graphs, pictographs, and/or pie charts) to reveal patterns that indicate relationships.	• Construct, analyze, and/or interpret graphical displays of data and/or large data sets to identify linear and nonlinear relationships. • Use graphical displays (e.g., maps, charts, graphs, and/or tables) of large data sets to identify temporal and spatial relationships. • Distinguish between causal and correlational relationships in data. • Analyze and interpret data to provide evidence for phenomena.	• Analyze data using tools, technologies, and/or models (e.g., computational, mathematical) in order to make valid and reliable scientific claims or determine an optimal design solution.
• N/A	• Analyze and interpret data to make sense of phenomena, using logical reasoning, mathematics, and/or computation.	• Apply concepts of statistics and probability (including mean, median, mode, and variability) to analyze and characterize data, using digital tools when feasible.	• Apply concepts of statistics and probability (including determining function fits to data, slope, intercept, and correlation coefficient for linear fits) to scientific and engineering questions and problems, using digital tools when feasible.
• N/A	• N/A	• Consider limitations of data analysis (e.g., measurement error) and/or seek to improve precision and accuracy of data with better technological tools and methods (e.g., multiple trials).	• Consider limitations of data analysis (e.g., measurement error, sample selection) when analyzing and interpreting data.
• N/A	• Compare and contrast data collected by different groups in order to discuss similarities and differences in their findings.	• Analyze and interpret data to determine similarities and differences in findings.	• Compare and contrast various types of data sets (e.g., self-generated, archival) to examine consistency of measurements and observations.
• Analyze data from tests of an object or tool to determine if it works as intended.	• Analyze data to refine a problem statement or the design of a proposed object, tool, or process. • Use data to evaluate and refine design solutions.	• Analyze data to define an optimal operational range for a proposed object, tool, process, or system that best meets criteria for success.	• Evaluate the impact of new data on a working explanation and/or model of a proposed process or system. • Analyze data to identify design features or characteristics of the components of a proposed process or system to optimize it relative to criteria for success.

N/A = Not applicable for this grade range

Science and Engineering Practices: Using Mathematics and Computational Thinking

In both science and engineering, mathematics and computation are fundamental tools for representing physical variables and their relationships. They are used for a range of tasks such as constructing simulations; solving equations exactly or approximately; and recognizing, expressing, and applying quantitative relationships. Mathematical and computational approaches enable scientists and engineers to predict the behavior of systems and test the validity of such predictions.

K–2 Condensed Practices	3–5 Condensed Practices	6–8 Condensed Practices	9–12 Condensed Practices
Mathematical and computational thinking in K–2 builds on prior experience and progresses to recognizing that mathematics can be used to describe the natural and designed world(s).	Mathematical and computational thinking in 3–5 builds on K–2 experiences and progresses to extending quantitative measurements to a variety of physical properties and using computation and mathematics to analyze data and compare alternative design solutions.	Mathematical and computational thinking in 6–8 builds on K–5 experiences and progresses to identifying patterns in large data sets and using mathematical concepts to support explanations and arguments.	Mathematical and computational thinking in 9–12 builds on K–8 experiences and progresses to using algebraic thinking and analysis, a range of linear and nonlinear functions including trigonometric functions, exponentials and logarithms, and computational tools for statistical analysis to analyze, represent, and model data. Simple computational simulations are created and used based on mathematical models of basic assumptions.
• N/A	• N/A	• Decide when to use qualitative vs. quantitative data.	• Decide if qualitative or quantitative data are best to determine whether a proposed object or tool meets criteria for success.
• Use counting and numbers to identify and describe patterns in the natural and designed world(s).	• Organize simple data sets to reveal patterns that suggest relationships.	• Use digital tools (e.g., computers) to analyze very large data sets for patterns and trends.	• Create and/or revise a computational model or simulation of a phenomenon, designed device, process, or system.
• Describe, measure, and/or compare quantitative attributes of different objects and display the data using simple graphs.	• Describe, measure, estimate, and/or graph quantities such as area, volume, weight, and time to address scientific and engineering questions and problems.	• Use mathematical representations to describe and/or support scientific conclusions and design solutions.	• Use mathematical, computational, and/or algorithmic representations of phenomena or design solutions to describe and/or support claims and/or explanations.
• Use quantitative data to compare two alternative solutions to a problem.	• Create and/or use graphs and/or charts generated from simple algorithms to compare alternative solutions to an engineering problem.	• Create algorithms (a series of ordered steps) to solve a problem. • Apply mathematical concepts and/or processes (such as ratio, rate, percent, basic operations, and simple algebra) to scientific and engineering questions and problems. • Use digital tools and/or mathematical concepts and arguments to test and compare proposed solutions to an engineering design problem.	• Apply techniques of algebra and functions to represent and solve scientific and engineering problems. • Use simple limit cases to test mathematical expressions, computer programs, algorithms, or simulations of a process or system to see if a model "makes sense" by comparing the outcomes with what is known about the real world. • Apply ratios, rates, percentages, and unit conversions in the context of complicated measurement problems involving quantities with derived or compound units (e.g., mg/mL, kg/m^3, acre-feet).

N/A = Not applicable for this grade range

Science and Engineering Practices: Constructing Explanations and Designing Solutions

The end-products of science are explanations and the end-products of engineering are solutions. The goal of science is the construction of theories that provide explanatory accounts of the world. A theory becomes accepted when it has multiple lines of empirical evidence and greater explanatory power of phenomena than previous theories. The goal of engineering design is to find a systematic solution to problems that is based on scientific knowledge and models of the material world. Each proposed solution results from a process of balancing competing criteria of desired functions, technical feasibility, cost, safety, aesthetics, and compliance with legal requirements. The optimal choice depends on how well the proposed solutions meet criteria and constraints.

K–2 Condensed Practices	3–5 Condensed Practices	6–8 Condensed Practices	9–12 Condensed Practices
Constructing explanations and designing solutions in K–2 builds on prior experiences and progresses to the use of evidence and ideas in constructing evidence-based accounts of natural phenomena and designing solutions.	Constructing explanations and designing solutions in 3–5 builds on K–2 experiences and progresses to the use of evidence in constructing explanations that specify variables that describe and predict phenomena and in designing multiple solutions to design problems.	Constructing explanations and designing solutions in 6–8 builds on K–5 experiences and progresses to include constructing explanations and designing solutions supported by multiple sources of evidence consistent with scientific ideas, principles, and theories.	Constructing explanations and designing solutions in 9–12 builds on K–8 experiences and progresses to explanations and designs that are supported by multiple and independent student-generated sources of evidence consistent with scientific ideas, principles, and theories.
• Use information from observations (firsthand and from media) to construct an evidence-based account for natural phenomena.	• Construct an explanation of observed relationships (e.g., the distribution of plants in the backyard).	• Construct an explanation that includes qualitative or quantitative relationships between variables that predict and/or describe phenomena. • Construct an explanation using models or representations.	• Make a quantitative and/or qualitative claim regarding the relationship between dependent and independent variables.
• N/A	• Use evidence (e.g., measurements, observations, patterns) to construct or support an explanation or design a solution to a problem.	• Construct a scientific explanation based on valid and reliable evidence obtained from sources (including the students' own experiments) and the assumption that theories and laws that describe the natural world operate today as they did in the past and will continue to do so in the future. • Apply scientific ideas, principles, and/or evidence to construct, revise and/or use an explanation for real-world phenomena, examples, or events.	• Construct and revise an explanation based on valid and reliable evidence obtained from a variety of sources (including students' own investigations, models, theories, simulations, peer review) and the assumption that theories and laws that describe the natural world operate today as they did in the past and will continue to do so in the future. • Apply scientific ideas, principles, and/or evidence to provide an explanation of phenomena and solve design problems, taking into account possible unanticipated effects.
• N/A	• Identify the evidence that supports particular points in an explanation.	• Apply scientific reasoning to show why the data or evidence is adequate for the explanation or conclusion.	• Apply scientific reasoning, theory, and/or models to link evidence to the claims to assess the extent to which the reasoning and data support the explanation or conclusion.
• Use tools and/or materials to design and/or build a device that solves a specific problem or a solution to a specific problem. • Generate and/or compare multiple solutions to a problem.	• Apply scientific ideas to solve design problems. • Generate and compare multiple solutions to a problem based on how well they meet the criteria and constraints of the design solution.	• Apply scientific ideas or principles to design, construct, and/or test a design of an object, tool, process, or system. • Undertake a design project, engaging in the design cycle, to construct and/or implement a solution that meets specific design criteria and constraints. • Optimize performance of a design by prioritizing criteria, making trade-offs, testing, revising, and retesting.	• Design, evaluate, and/or refine a solution to a complex real-world problem, based on scientific knowledge, student-generated sources of evidence, prioritized criteria, and trade-off considerations.

N/A = Not applicable for this grade range

Science and Engineering Practices: Engaging in Argument From Evidence

Argumentation is the process by which evidence-based conclusions and solutions are reached. In science and engineering, reasoning and argument based on evidence are essential to identifying the best explanation for a natural phenomenon or the best solution to a design problem. Scientists and engineers use argumentation to listen to, compare, and evaluate competing ideas and methods based on merits. Scientists and engineers engage in argumentation when investigating a phenomenon, testing a design solution, resolving questions about measurements, building data models, and using evidence to evaluate claims.

K–2 Condensed Practices	3–5 Condensed Practices	6–8 Condensed Practices	9–12 Condensed Practices
Engaging in argument from evidence in K–2 builds on prior experiences and progresses to comparing ideas and representations about the natural and designed world(s).	Engaging in argument from evidence in 3–5 builds on K–2 experiences and progresses to critiquing the scientific explanations or solutions proposed by peers by citing relevant evidence about the natural and designed world(s).	Engaging in argument from evidence in 6–8 builds on K–5 experiences and progresses to constructing a convincing argument that supports or refutes claims for either explanations or solutions about the natural and designed world(s).	Engaging in argument from evidence in 9–12 builds on K–8 experiences and progresses to using appropriate and sufficient evidence and scientific reasoning to defend and critique claims and explanations about the natural and designed world(s). Arguments may also come from current scientific or historical episodes in science.
• Identify arguments that are supported by evidence. • Distinguish between explanations that account for all gathered evidence and those that do not. • Analyze why some evidence is relevant to a scientific question and some is not. • Distinguish between opinions and evidence in one's own explanations.	• Compare and refine arguments based on an evaluation of the evidence presented. • Distinguish among facts, reasoned judgment based on research findings, and speculation in an explanation.	• Compare and critique two arguments on the same topic and analyze whether they emphasize similar or different evidence and/or interpretations of facts.	• Compare and evaluate competing arguments or design solutions in light of currently accepted explanations, new evidence, limitations (e.g., trade-offs), constraints, and ethical issues. • Evaluate the claims, evidence, and/or reasoning behind currently accepted explanations or solutions to determine the merits of arguments.
• Listen actively to arguments to indicate agreement or disagreement based on evidence, and/or to retell the main points of the argument.	• Respectfully provide and receive critiques from peers about a proposed procedure, explanation, or model by citing relevant evidence and posing specific questions.	• Respectfully provide and receive critiques about one's explanations, procedures, models, and questions by citing relevant evidence and posing and responding to questions that elicit pertinent elaboration and detail.	• Respectfully provide and/or receive critiques on scientific arguments by probing reasoning and evidence and challenging ideas and conclusions, responding thoughtfully to diverse perspectives, and determining what additional information is required to resolve contradictions.
• Construct an argument with evidence to support a claim.	• Construct and/or support an argument with evidence, data, and/or a model. • Use data to evaluate claims about cause and effect.	• Construct, use, and/or present an oral and written argument supported by empirical evidence and scientific reasoning to support or refute an explanation or a model for a phenomenon or a solution to a problem.	• Construct, use, and/or present an oral and written argument or counter-arguments based on data and evidence.
• Make a claim about the effectiveness of an object, tool, or solution that is supported by relevant evidence.	• Make a claim about the merit of a solution to a problem by citing relevant evidence about how it meets the criteria and constraints of the problem.	• Make an oral or written argument that supports or refutes the advertised performance of a device, process, or system, based on empirical evidence concerning whether or not the technology meets relevant criteria and constraints. • Evaluate competing design solutions based on jointly developed and agreed-upon design criteria.	• Make and defend a claim based on evidence about the natural world or the effectiveness of a design solution that reflects scientific knowledge and student-generated evidence. • Evaluate competing design solutions to a real-world problem based on scientific ideas and principles, empirical evidence, and/or logical arguments regarding relevant factors (e.g., economic, societal, environmental, ethical considerations).

Science and Engineering Practices: Obtaining, Evaluating, and Communicating Information

Scientists and engineers must be able to communicate clearly and persuasively the ideas and methods they generate. Critiquing and communicating ideas individually and in groups is a critical professional activity. Communicating information and ideas can be done in multiple ways: using tables, diagrams, graphs, models, and equations as well as orally, in writing, and through extended discussions. Scientists and engineers employ multiple sources to obtain information that is used to evaluate the merit and validity of claims, methods, and designs.

K–2 Condensed Practices	3–5 Condensed Practices	6–8 Condensed Practices	9–12 Condensed Practices
Obtaining, evaluating, and communicating information in K–2 builds on prior experiences and uses observations and texts to communicate new information.	Obtaining, evaluating, and communicating information in 3–5 builds on K–2 experiences and progresses to evaluating the merit and accuracy of ideas and methods.	Obtaining, evaluating, and communicating information in 6–8 builds on K–5 experiences and progresses to evaluating the merit and validity of ideas and methods.	Obtaining, evaluating, and communicating information in 9–12 builds on K–8 experiences and progresses to evaluating the validity and reliability of the claims, methods, and designs.
• Read grade-appropriate texts and/or use media to obtain scientific and/or technical information to determine patterns in and/or evidence about the natural and designed world(s).	• Read and comprehend grade-appropriate complex texts and/or other reliable media to summarize and obtain scientific and technical ideas and describe how they are supported by evidence. • Compare and/or combine across complex texts and/or other reliable media to support the engagement in other scientific and/or engineering practices.	• Critically read scientific texts adapted for classroom use to determine the central ideas and/or obtain scientific and/or technical information to describe patterns in and/or evidence about the natural and designed world(s).	• Critically read scientific literature adapted for classroom use to determine the central ideas or conclusions and/or to obtain scientific and/or technical information to summarize complex evidence, concepts, processes, or information presented in a text by paraphrasing them in simpler but still accurate terms.
• Describe how specific images (e.g., a diagram showing how a machine works) support a scientific or engineering idea.	• Combine information in written text with that contained in corresponding tables, diagrams, and/or charts to support the engagement in other scientific and/or engineering practices.	• Integrate qualitative and/or quantitative scientific and/or technical information in written text with that contained in media and visual displays to clarify claims and findings.	• Compare, integrate, and evaluate sources of information presented in different media or formats (e.g., visually, quantitatively) as well as in words in order to address a scientific question or solve a problem.
• Obtain information using various texts, text features (e.g., headings, tables of contents, glossaries, electronic menus, icons), and other media that will be useful in answering a scientific question and/or supporting a scientific claim.	• Obtain and combine information from books and/or other reliable media to explain phenomena or solutions to a design problem.	• Gather, read, and synthesize information from multiple appropriate sources and assess the credibility, accuracy, and possible bias of each publication and methods used, and describe how they are supported or not supported by evidence. • Evaluate data, hypotheses, and/or conclusions in scientific and technical texts in light of competing information or accounts.	• Gather, read, and evaluate scientific and/or technical information from multiple authoritative sources, assessing the evidence and usefulness of each source. • Evaluate the validity and reliability of and/or synthesize multiple claims, methods, and/or designs that appear in scientific and technical texts or media reports, verifying the data when possible.
• Communicate information or design ideas and/or solutions with others in oral and/or written forms using models, drawings, writing, or numbers that provide detail about scientific ideas, practices, and/or design ideas.	• Communicate scientific and/or technical information orally and/or in written formats, including various forms of media and may include tables, diagrams, and charts.	• Communicate scientific and/or technical information (e.g., about a proposed object, tool, process, system) in writing and/or through oral presentations.	• Communicate scientific and/or technical information or ideas (e.g., about phenomena and/or the process of development and the design and performance of a proposed process or system) in multiple formats (including orally, graphically, textually, and mathematically).

Crosscutting Concepts

Grades K–2	Grades 3–5	Grades 6–8	Grades 9–12
Patterns: Observed patterns in nature guide organization and classification and prompt questions about relationships and causes underlying them.			
• Patterns in the natural and human designed world can be observed, used to describe phenomena, and used as evidence.	• Similarities and differences in patterns can be used to sort, classify, communicate, and analyze simple rates of change for natural phenomena and designed products. • Patterns of change can be used to make predictions. • Patterns can be used as evidence to support an explanation.	• Macroscopic patterns are related to the nature of microscopic and atomic-level structure. • Patterns in rates of change and other numerical relationships can provide information about natural and human-designed systems. • Patterns can be used to identify cause-and-effect relationships. • Graphs, charts, and images can be used to identify patterns in data.	• Different patterns may be observed at each of the scales at which a system is studied and can provide evidence for causality in explanations of phenomena. • Classifications or explanations used at one scale may fail or need revision when information from smaller or larger scales is introduced, thus requiring improved investigations and experiments. • Patterns of performance of designed systems can be analyzed and interpreted to reengineer and improve the system. • Mathematical representations are needed to identify some patterns. • Empirical evidence is needed to identify patterns.
Cause and Effect: Mechanism and Prediction: Events have causes, sometimes simple, sometimes multifaceted. Deciphering causal relationships, and the mechanisms by which they are mediated, is a major activity of science and engineering.			
• Events have causes that generate observable patterns. • Simple tests can be designed to gather evidence to support or refute student ideas about causes.	• Cause-and-effect relationships are routinely identified, tested, and used to explain change. • Events that occur together with regularity might or might not be a cause-and-effect relationship.	• Relationships can be classified as causal or correlational, and correlation does not necessarily imply causation. • Cause-and-effect relationships may be used to predict phenomena in natural or designed systems. • Phenomena may have more than one cause, and some cause-and-effect relationships in systems can only be described using probability.	• Empirical evidence is required to differentiate between cause and correlation and make claims about specific causes and effects. • Cause-and-effect relationships can be suggested and predicted for complex natural and human-designed systems by examining what is known about smaller scale mechanisms within the system. • Systems can be designed to cause a desired effect. • Changes in systems may have various causes that may not have equal effects.

Crosscutting Concepts (*continued*)

Grades K–2	Grades 3–5	Grades 6–8	Grades 9–12
Scale, Proportion, and Quantity: In considering phenomena, it is critical to recognize what is relevant at different size, time, and energy scales, and to recognize proportional relationships between different quantities as scales change.			
• Relative scales allow objects and events to be compared and described (e.g., bigger and smaller; hotter and colder; faster and slower). • Standard units are used to measure length.	• Natural objects and/or observable phenomena exist from the very small to the immensely large or from very short to very long time periods. • Standard units are used to measure and describe physical quantities such as weight, time, temperature, and volume.	• Time, space, and energy phenomena can be observed at various scales using models to study systems that are too large or too small. • The observed function of natural and designed systems may change with scale. • Proportional relationships (e.g., speed as the ratio of distance traveled to time taken) among different types of quantities provide information about the magnitude of properties and processes. • Scientific relationships can be represented through the use of algebraic expressions and equations. • Phenomena that can be observed at one scale may not be observable at another scale.	• The significance of a phenomenon is dependent on the scale, proportion, and quantity at which it occurs. • Some systems can only be studied indirectly because they are too small, too large, too fast, or too slow to observe directly. • Patterns observable at one scale may not be observable or exist at other scales. • Using the concept of orders of magnitude allows one to understand how a model at one scale relates to a model at another scale. • Algebraic thinking is used to examine scientific data and predict the effect of a change in one variable on another (e.g., linear growth vs. exponential growth).
Systems and System Models: A system is an organized group of related objects or components; models can be used for understanding and predicting the behavior of systems.			
• Objects and organisms can be described in terms of their parts. • Systems in the natural and designed world have parts that work together.	• A system is a group of related parts that make up a whole and can carry out functions its individual parts cannot. • A system can be described in terms of its components and their interactions.	• Systems may interact with other systems; they may have subsystems and be a part of larger complex systems. • Models can be used to represent systems and their interactions—such as inputs, processes, and outputs—and energy, matter, and information flows within systems. • Models are limited in that they only represent certain aspects of the system under study.	• Systems can be designed to do specific tasks. • When investigating or describing a system, the boundaries and initial conditions of the system need to be defined and their inputs and outputs analyzed and described using models. • Models (e.g., physical, mathematical, computer models) can be used to simulate systems and interactions—including energy, matter, and information flows—within and between systems at different scales. • Models can be used to predict the behavior of a system, but these predictions have limited precision and reliability due to the assumptions and approximations inherent in models.

Crosscutting Concepts *(continued)*

Grades K–2	Grades 3–5	Grades 6–8	Grades 9–12
Energy and Matter: Flows, Cycles, and Conservation: Tracking energy and matter flows into, out of, and within systems helps one understand their system's behavior.			
• Objects may break into smaller pieces, be put together into larger pieces, or change shapes.	• Matter is made of particles. • Matter flows and cycles can be tracked in terms of the weight of the substances before and after a process occurs. The total weight of the substances does not change. This is what is meant by conservation of matter. Matter is transported into, out of, and within systems. • Energy can be transferred in various ways and between objects.	• Matter is conserved because atoms are conserved in physical and chemical processes. • Within a natural or designed system, the transfer of energy drives the motion and/or cycling of matter. • Energy may take different forms (e.g. energy in fields, thermal energy, energy of motion). • The transfer of energy can be tracked as energy flows through a designed or natural system.	• The total amount of energy and matter in closed systems is conserved. • Changes of energy and matter in a system can be described in terms of energy and matter flows into, out of, and within that system. • Energy cannot be created or destroyed—it only moves between one place and another place, between objects and/or fields, or between systems. • Energy drives the cycling of matter within and between systems. • In nuclear processes, atoms are not conserved, but the total number of protons plus neutrons is conserved.
Structure and Function: The way an object is shaped or structured determines many of its properties and functions.			
• The shape and stability of structures of natural and designed objects are related to their function(s).	• Different materials have different substructures, which can sometimes be observed. • Substructures have shapes and parts that serve functions.	• Complex and microscopic structures and systems can be visualized, modeled, and used to describe how their function depends on the shapes, composition, and relationships among its parts; therefore, complex natural and designed structures/systems can be analyzed to determine how they function. • Structures can be designed to serve particular functions by taking into account properties of different materials, and how materials can be shaped and used.	• Investigating or designing new systems or structures requires a detailed examination of the properties of different materials, the structures of different components, and connections of components to reveal their function and/or solve a problem. • The functions and properties of natural and designed objects and systems can be inferred from their overall structure, the way their components are shaped and used, and the molecular substructures of their various materials.
Stability and Change: For both designed and natural systems, conditions that affect stability and factors that control rates of change are critical elements to consider and understand.			
• Some things stay the same while other things change. • Things may change slowly or rapidly.	• Change is measured in terms of differences over time and may occur at different rates. • Some systems appear stable, but over long periods of time will eventually change.	• Explanations of stability and change in natural or designed systems can be constructed by examining the changes over time and processes at different scales, including the atomic scale. • Small changes in one part of a system might cause large changes in another part. • Stability might be disturbed by either sudden events or gradual changes that accumulate over time. • Systems in dynamic equilibrium are stable due to a balance of feedback mechanisms.	• Much of science deals with constructing explanations of how things change and how they remain stable. • Change and rates of change can be quantified and modeled over very short or very long periods of time. Some system changes are irreversible. • Feedback (negative or positive) can stabilize or destabilize a system. • Systems can be designed for greater or lesser stability.

Disciplinary Core Ideas in Physical Science

	Grades K–2	Grades 3–5	Grades 6–8	Grades 9–12
PS1: Matter and Its Interactions				
PS1.A: Structure and Properties of Matter	• Different kinds of matter exist and many of them can be either solid or liquid, depending on temperature. Matter can be described and classified by its observable properties. (2-PS1-1) • Different properties are suited to different purposes. (2-PS1-2),(2-PS1-3) • A great variety of objects can be built up from a small set of pieces. (2-PS1-3)	• Matter of any type can be subdivided into particles that are too small to see, but even then the matter still exists and can be detected by other means. A model shows that gases are made from matter particles that are too small to see and are moving freely around in space can explain many observations, including the inflation and shape of a balloon and the effects of air on larger particles or objects. (5-PS1-1) • The amount (weight) of matter is conserved when it changes form, even in transitions in which it seems to vanish. (5-PS1-2) • Measurements of a variety of properties can be used to identify materials. (Boundary: At this grade level, mass and weight are not distinguished, and no attempt is made to define the unseen particles or explain the atomic-scale mechanism of evaporation and condensation.) (5-PS1-3)	• Substances are made from different types of atoms, which combine with one another in various ways. Atoms form molecules that range in size from two to thousands of atoms. (MS-PS1-1) • Each pure substance has characteristic physical and chemical properties (for any bulk quantity under given conditions) that can be used to identify it. (MS-PS1-2), (MS-PS1-3) • Gases and liquids are made of molecules or inert atoms that are moving about relative to each other. (MS-PS1-4) • In a liquid, the molecules are constantly in contact with others; in a gas, they are widely spaced except when they happen to collide. In a solid, atoms are closely spaced and may vibrate in position but do not change relative locations. (MS-PS1-4) • Solids may be formed from molecules, or they may be extended structures with repeating subunits (e.g., crystals). (MS-PS1-1) • The changes of state that occur with variations in temperature or pressure can be described and predicted using these models of matter. (MS-PS1-4)	• Each atom has a charged substructure consisting of a nucleus, which is made of protons and neutrons, surrounded by electrons. (HS-PS1-1) • The periodic table orders elements horizontally by the number of protons in the atom's nucleus and places those with similar chemical properties in columns. The repeating patterns of this table reflect patterns of outer electron states. (HS-PS1-1),(HS-PS1-2) • The structure and interactions of matter at the bulk scale are determined by electrical forces within and between atoms. (HS-PS1-3),(secondary to HS-PS2-6) • Stable forms of matter are those in which the electric and magnetic field energy is minimized. A stable molecule has less energy than the same set of atoms separated; one must provide at least this energy in order to take the molecule apart. (HS-PS1-4)

Disciplinary Core Ideas in Physical Science (*continued*)

	Grades K–2	Grades 3–5	Grades 6–8	Grades 9–12
PS1.B: Chemical Reactions	• Heating or cooling a substance may cause changes that can be observed. Sometimes these changes are reversible, and sometimes they are not. (2-PS1-4)	• When two or more different substances are mixed, a new substance with different properties may be formed. (5-PS1-4) • No matter what reaction or change in properties occurs, the total weight of the substances does not change. (Boundary: Mass and weight are not distinguished at this grade level.) (5-PS1-2)	• Substances react chemically in characteristic ways. In a chemical process, the atoms that make up the original substances are regrouped into different molecules, and these new substances have different properties from those of the reactants. (MS-PS1-2),(MS-PS1-3),(MS-PS1-5) • The total number of each type of atom is conserved, and thus the mass does not change. (MS-PS1-5) • Some chemical reactions release energy, others store energy. (MS-PS1-6)	• Chemical processes, their rates, and whether or not energy is stored or released can be understood in terms of the collisions of molecules and the rearrangements of atoms into new molecules, with consequent changes in the sum of all bond energies in the set of molecules that are matched by changes in kinetic energy. (HS-PS1-4),(HS-PS1-5) • In many situations, a dynamic and condition-dependent balance between a reaction and the reverse reaction determines the numbers of all types of molecules present. (HS-PS1-6) • The fact that atoms are conserved, together with knowledge of the chemical properties of the elements involved, can be used to describe and predict chemical reactions. (HS-PS1-2),(HS-PS1-7)
PS1.C: Nuclear Processes	• N/A	• N/A	• N/A	• Nuclear processes, including fusion, fission, and radioactive decays of unstable nuclei, involve release or absorption of energy. The total number of neutrons plus protons does not change in any nuclear process. (HS-PS1-8) • Spontaneous radioactive decays follow a characteristic exponential decay law. Nuclear lifetimes allow radiometric dating to be used to determine the ages of rocks and other materials. (secondary to HS-ESS1-5),(secondary to HS-ESS1-6)

N/A = Not applicable for this grade range

Disciplinary Core Ideas in Physical Science (*continued*)

	Grades K–2	Grades 3–5	Grades 6–8	Grades 9–12
PS2: Motion and Stability: Forces and Interactions				
PS2.A: Forces and Motion	• Pushes and pulls can have different strengths and directions. (K-PS2-1),(K-PS2-2) • Pushing or pulling on an object can change the speed or direction of its motion and can start or stop it. (K-PS2-1),(K-PS2-2)	• Each force acts on one particular object and has both strength and a direction. An object at rest typically has multiple forces acting on it, but they add to give zero net force on the object. Forces that do not sum to zero can cause changes in the object's speed or direction of motion. (Boundary: Qualitative and conceptual, but not quantitative, addition of forces are used at this level.) (3-PS2-1) • The patterns of an object's motion in various situations can be observed and measured; when that past motion exhibits a regular pattern, future motion can be predicted from it. (Boundary: Technical terms, such as magnitude, velocity, momentum, and vector quantity, are not introduced at this level, but the concept that some quantities need both size and direction to be described is developed.) (3-PS2-2)	• For any pair of interacting objects, the force exerted by the first object on the second object is equal in strength to the force that the second object exerts on the first, but in the opposite direction (Newton's third law). (MS-PS2-1) • The motion of an object is determined by the sum of the forces acting on it; if the total force on the object is not zero, its motion will change. The greater the mass of the object, the greater the force needed to achieve the same change in motion. For any given object, a larger force causes a larger change in motion. (MS-PS2-2) • All positions of objects and the directions of forces and motions must be described in an arbitrarily chosen reference frame and arbitrarily chosen units of size. In order to share information with other people, these choices must also be shared. (MSPS2-2)	• Newton's second law accurately predicts changes in the motion of macroscopic objects. (HS-PS2-1) • Momentum is defined for a particular frame of reference; it is the mass times the velocity of the object. In any system, total momentum is always conserved. (HS-PS2-2) • If a system interacts with objects outside itself, the total momentum of the system can change; however, any such change is balanced by changes in the momentum of objects outside the system. (HS-PS2-2),(HS-PS2-3)
PS2.B: Types of Interactions	• When objects touch or collide, they push on one another and can change motion. (K-PS2-1)	• Objects in contact exert forces on each other. (3-PS2-1) • Electric and magnetic forces between a pair of objects do not require that the objects be in contact. The sizes of the forces in each situation depend on the properties of the objects and their distances apart and, for forces between two magnets, on their orientation relative to each other. (3-PS2-3),(3-PS2-4) • The gravitational force of Earth acting on an object near Earth's surface pulls that object toward the planet's center. (5-PS2-1)	• Electric and magnetic (electromagnetic) forces can be attractive or repulsive, and their sizes depend on the magnitudes of the charges, currents, or magnetic strengths involved and on the distances between the interacting objects. (MS-PS2-3) • Gravitational forces are always attractive. There is a gravitational force between any two masses, but it is very small except when one or both of the objects have large mass—e.g., Earth and the Sun. (MS-PS2-4) • Forces that act at a distance (electric, magnetic, and gravitational) can be explained by fields that extend through space and can be mapped by their effect on a test object (a charged object, a magnet, or a ball, respectively). (MS-PS2-5)	• Newton's law of universal gravitation and Coulomb's law provide the mathematical models to describe and predict the effects of gravitational and electrostatic forces between distant objects. (HS-PS2-4) • Forces at a distance are explained by fields (gravitational, electric, and magnetic) permeating space that can transfer energy through space. Magnets or electric currents cause magnetic fields; electric charges or changing magnetic fields cause electric fields. (HS-PS2-4),(HS-PS2-5) • Attraction and repulsion between electric charges at the atomic scale explain the structure, properties, and transformations of matter, as well as the contact forces between material objects. (HS-PS2-6),(secondary to HS-PS1-1),(secondary to HS-PS1-3)

Disciplinary Core Ideas in Physical Science (*continued*)

	Grades K–2	Grades 3–5	Grades 6–8	Grades 9–12
PS3: Energy				
PS3.A: Definitions of Energy	• N/A	• The faster a given object is moving, the more energy it possesses. (4-PS3-1) • Energy can be moved from place to place by moving objects or through sound, light, or electric currents. (4-PS3-2),(4-PS3-3)	• Motion energy is properly called kinetic energy; it is proportional to the mass of the moving object and grows with the square of its speed. (MS-PS3-1) • A system of objects may also contain stored (potential) energy, depending on their relative positions. (MS-PS3-2) • The term "heat" as used in everyday language refers both to thermal energy (the motion of atoms or molecules within a substance) and the transfer of that thermal energy from one object to another. In science, heat is used only for this second meaning; it refers to the energy transferred due to the temperature difference between two objects.(secondary to MS-PS1-4) • Temperature is not a measure of energy; the relationship between the temperature and the total energy of a system depends on the types, states, and amounts of matter present. (secondary to MS-PS1-4)	• Energy is a quantitative property of a system that depends on the motion and interactions of matter and radiation within that system. That there is a single quantity called energy is due to the fact that a system's total energy is conserved, even as, within the system, energy is continually transferred from one object to another and between its various possible forms. (HS-PS3-1),(HS-PS3-2) • At the macroscopic scale, energy manifests itself in multiple ways, such as in motion, sound, light, and thermal energy. (HS-PS3-2), (HS-PS3-3) • "Electrical energy" may mean energy stored in a battery or energy transmitted by electric currents. (secondary to HS-PS2-5) • These relationships are better understood at the microscopic scale, at which all of the different manifestations of energy can be modeled as a combination of energy associated with the motion of particles and energy associated with the configuration (relative position of the particles). In some cases the relative position energy can be thought of as stored in fields (which mediate interactions between particles). This last concept includes radiation, a phenomenon in which energy stored in fields moves across space. (HS-PS3-2)

N/A = Not applicable for this grade range

Disciplinary Core Ideas in Physical Science (*continued*)

	Grades K–2	Grades 3–5	Grades 6–8	Grades 9–12
PS3.B: Conservation of Energy and Energy Transfer	• Sunlight warms Earth's surface. (K-PS3-1),(K-PS3-2)	• Energy is present whenever there are moving objects, sound, light, or heat. When objects collide, energy can be transferred from one object to another, thereby changing their motion. In such collisions, some energy is typically also transferred to the surrounding air; as a result, the air gets heated and sound is produced. (4-PS3-2),(4-PS3-3) • Light also transfers energy from place to place. (4-PS3-2) • Energy can also be transferred from place to place by electric currents, which can then be used locally to produce motion, sound, heat, or light. The currents may have been produced to begin with by transforming the energy of motion into electrical energy. (4-PS3-2),(4-PS3-4)	• When the motion energy of an object changes, there is inevitably some other change in energy at the same time. (MS-PS3-5) • The amount of energy transfer needed to change the temperature of a matter sample by a given amount depends on the nature of the matter, the size of the sample, and the environment. (MS-PS3-4) • Energy is spontaneously transferred out of hotter regions or objects and into colder ones. (MS-PS3-3)	• Conservation of energy means that the total change of energy in any system is always equal to the total energy transferred into or out of the system. (HS-PS3-1) • Energy cannot be created or destroyed, but it can be transported from one place to another and transferred between systems. (HS-PS3-1),(HS-PS3-4) • Mathematical expressions, which quantify how the stored energy in a system depends on its configuration (e.g., relative positions of charged particles, compression of a spring) and how kinetic energy depends on mass and speed, allow the concept of conservation of energy to be used to predict and describe system behavior. (HS-PS3-1) • The availability of energy limits what can occur in any system. (HS-PS3-1) • Uncontrolled systems always evolve toward more stable states—that is, toward more uniform energy distribution (e.g., water flows downhill, objects hotter than their surrounding environment cool down). (HS-PS3-4)
PS3.C: Relationship Between Energy and Forces	• A bigger push or pull makes things go faster. (secondary to K-PS2-1)	• When objects collide, the contact forces transfer energy so as to change the objects' motions. (4-PS3-3)	• When two objects interact, each one exerts a force on the other that can cause energy to be transferred to or from the object. (MS-PS3-2)	• When two objects interacting through a field change relative position, the energy stored in the field is changed. (HS-PS3-5)

Disciplinary Core Ideas in Physical Science (*continued*)

	Grades K–2	Grades 3–5	Grades 6–8	Grades 9–12
PS3.D: Energy in Chemical Processes and Everyday Life	• N/A	• The expression "produce energy" typically refers to the conversion of stored energy into a desired form for practical use. (4-PS3-4) • The energy released [from] food was once energy from the Sun that was captured by plants in the chemical process that forms plant matter (from air and water). (5-PS3-1)	• The chemical reaction by which plants produce complex food molecules (sugars) requires an energy input (i.e., from sunlight) to occur. In this reaction, carbon dioxide and water combine to form carbon-based organic molecules and release oxygen. (secondary to MS-LS1-6) • Cellular respiration in plants and animals involve chemical reactions with oxygen that release stored energy. In these processes, complex molecules containing carbon react with oxygen to produce carbon dioxide and other materials. (secondary to MS-LS1-7)	• Although energy cannot be destroyed, it can be converted to less useful forms—for example, to thermal energy in the surrounding environment. (HS-PS3-3),(HS-PS3-4) • Solar cells are human-made devices that likewise capture the Sun's energy and produce electrical energy. (secondary to HS-PS4-5) • The main way that solar energy is captured and stored on Earth is through the complex chemical process known as photosynthesis. (secondary to HS-LS2-5) • Nuclear fusion processes in the center of the Sun release the energy that ultimately reaches Earth as radiation. (secondary to HS-ESS1-1)
PS4: Waves and Their Applications in Technologies for Information Transfer				
PS4.A: Wave Properties	• Sound can make matter vibrate, and vibrating matter can make sound. (1-PS4-1)	• Waves, which are regular patterns of motion, can be made in water by disturbing the surface. When waves move across the surface of deep water, the water goes up and down in place; there is no net motion in the direction of the wave except when the water meets a beach. (Note: This grade band endpoint was moved from K–2.) (4-PS4-1) • Waves of the same type can differ in amplitude (height of the wave) and wavelength (spacing between wave peaks). (4-PS4-1)	• A simple wave has a repeating pattern with a specific wavelength, frequency, and amplitude. (MS-PS4-1) • A sound wave needs a medium through which it is transmitted. (MS-PS4-2)	• The wavelength and frequency of a wave are related to one another by the speed of travel of the wave, which depends on the type of wave and the medium through which it is passing. (HS-PS4-1) • Information can be digitized (e.g., a picture stored as the values of an array of pixels); in this form, it can be stored reliably in computer memory and sent over long distances as a series of wave pulses. (HS-PS4-2),(HS-PS4-5) • Waves can add or cancel one another as they cross, depending on their relative phase (i.e., relative position of peaks and troughs of the waves), but they emerge unaffected by each other. (Boundary: The discussion at this grade level is qualitative only; it can be based on the fact that two different sounds can pass a location in different directions without getting mixed up.) (HS-PS4-3) (*Note:* This grade band endpoint was moved from 3–5.) • Geologists use seismic waves and their reflection at interfaces between layers to probe structures deep in the planet. (secondary to HS-ESS2-3)

N/A = Not applicable for this grade range

Disciplinary Core Ideas in Physical Science (*continued*)

	Grades K–2	Grades 3–5	Grades 6–8	Grades 9–12
PS4.B: Electromagnetic Radiation	• Objects can be seen only when light is available to illuminate them. Some objects give off their own light. (1-PS4-2) • Some materials allow light to pass through them, others allow only some light through, and still others block all the light and create a dark shadow on any surface beyond them, where the light cannot reach. Mirrors can be used to redirect a light beam. (Boundary: The idea that light travels from place to place is developed through experiences with light sources, mirrors, and shadows, but no attempt is made to discuss the speed of light.) (1-PS4-3)	• An object can be seen when light reflected from its surface enters the eyes. (4-PS4-2)	• When light shines on an object, it is reflected, absorbed, or transmitted through the object, depending on the object's material and the frequency (color) of the light. (MS-PS4-2) • The path that light travels can be traced as straight lines, except at surfaces between different transparent materials (e.g., air and water, air and glass) where the light path bends. (MS-PS4-2) • A wave model of light is useful for explaining brightness, color, and the frequency-dependent bending of light at a surface between media. (MS-PS4-2) • However, because light can travel through space, it cannot be a matter wave, like sound or water waves. (MS-PS4-2)	• Electromagnetic radiation (e.g., radio, microwaves, light) can be modeled as a wave of changing electric and magnetic fields or as particles called photons. The wave model is useful for explaining many features of electromagnetic radiation, and the particle model explains other features. (HS-PS4-3) • When light or longer wavelength electromagnetic radiation is absorbed in matter, it is generally converted into thermal energy (heat). Shorter wavelength electromagnetic radiation (ultraviolet, X-rays, gamma rays) can ionize atoms and cause damage to living cells. (HS-PS4-4) • Photovoltaic materials emit electrons when they absorb light of a high-enough frequency. (HS-PS4-5) • Atoms of each element emit and absorb characteristic frequencies of light. These characteristics allow identification of the presence of an element, even in microscopic quantities. (secondary to HS-ESS1-2)
PS4.C: Information Technologies and Instrumentation	• People use a variety of devices to communicate (send and receive information) over long distances. (1-PS4-4)	• Digitized information can be transmitted over long distances without significant degradation. High-tech devices, such as computers or cell phones, can receive and decode information—convert it from digitized form to voice—and vice versa. (4-PS4-3)	• Digitized signals (sent as wave pulses) are a more reliable way to encode and transmit information. (MS-PS4-3)	• Multiple technologies based on the understanding of waves and their interactions with matter are part of everyday experiences in the modern world (e.g., medical imaging, communications, scanners) and in scientific research. They are essential tools for producing, transmitting, and capturing signals and for storing and interpreting the information contained in them. (HS-PS4-5)

Disciplinary Core Ideas in Life Science

	Grades K–2	Grades 3–5	Grades 6–8	Grades 9–12
LS1: From Molecules to Organisms: Structures and Processes				
LS1.A: Structure and Function	• All organisms have external parts. Different animals use their body parts in different ways to see, hear, grasp objects, protect themselves, move from place to place, and seek, find, and take in food, water, and air. Plants also have different parts (roots, stems, leaves, flowers, fruits) that help them survive and grow. (1-LS1-1)	• Plants and animals have both internal and external structures that serve various functions in growth, survival, behavior, and reproduction. (4-LS1-1)	• All living things are made up of cells, which is the smallest unit that can be said to be alive. An organism may consist of one single cell (unicellular) or many different numbers and types of cells (multicellular). (MS-LS1-1) • Within cells, special structures are responsible for particular functions, and the cell membrane forms the boundary that controls what enters and leaves the cell. (MS-LS1-2) • In multicellular organisms, the body is a system of multiple interacting subsystems. These subsystems are groups of cells that work together to form tissues and organs that are specialized for particular body functions. (MS-LS1-3)	• Systems of specialized cells within organisms help them perform the essential functions of life. (HS-LS1-1) • All cells contain genetic information in the form of DNA molecules. Genes are regions in the DNA that contain the instructions that code for the formation of proteins, which carry out most of the work of cells. (HS-LS1-1) (secondary to HS-LS3-1) • Multicellular organisms have a hierarchical structural organization, in which any one system is made up of numerous parts and is itself a component of the next level. (HS-LS1-2) • Feedback mechanisms maintain a living system's internal conditions within certain limits and mediate behaviors, allowing it to remain alive and functional even as external conditions change within some range. Feedback mechanisms can encourage (through positive feedback) or discourage (negative feedback) what is going on inside the living system. (HS-LS1-3)
LS1.B: Growth and Development of Organisms	• Adult plants and animals can have young. In many kinds of animals, parents and the offspring themselves engage in behaviors that help the offspring to survive. (1-LS1-2)	• Reproduction is essential to the continued existence of every kind of organism. Plants and animals have unique and diverse life cycles. (3-LS1-1)	• Organisms reproduce, either sexually or asexually, and transfer their genetic information to their offspring. (secondary to MS-LS3-2) • Animals engage in characteristic behaviors that increase the odds of reproduction. (MS-LS1-4) • Plants reproduce in a variety of ways, sometimes depending on animal behavior and specialized features for reproduction. (MS-LS1-4) • Genetic factors as well as local conditions affect the growth of the adult plant. (MS-LS1-5)	• In multicellular organisms individual cells grow and then divide via a process called mitosis, thereby allowing the organism to grow. The organism begins as a single cell (fertilized egg) that divides successively to produce many cells, with each parent cell passing identical genetic material (two variants of each chromosome pair) to both daughter cells. Cellular division and differentiation produce and maintain a complex organism, composed of systems of tissues and organs that work together to meet the needs of the whole organism. (HS-LS1-4)

Disciplinary Core Ideas in Life Science (*continued*)

	Grades K–2	Grades 3–5	Grades 6–8	Grades 9–12
LS1.C: Organization for Matter and Energy Flow in Organisms	• All animals need food in order to live and grow. They obtain their food from plants or from other animals. Plants need water and light to live and grow. (K-LS1-1)	• Food provides animals with the materials they need for body repair and growth and the energy they need to maintain body warmth and for motion. (secondary to 5-PS3-1) • Plants acquire their material for growth chiefly from air and water. (5-LS1-1)	• Plants, algae (including phytoplankton), and many microorganisms use the energy from light to make sugars (food) from carbon dioxide from the atmosphere and water through the process of photosynthesis, which also releases oxygen. These sugars can be used immediately or stored for growth or later use. (MS-LS1-6) • Within individual organisms, food moves through a series of chemical reactions in which it is broken down and rearranged to form new molecules, to support growth, or to release energy. (MS-LS1-7)	• The process of photosynthesis converts light energy to stored chemical energy by converting carbon dioxide plus water into sugars plus released oxygen. (HS-LS1-5) • The sugar molecules thus formed contain carbon, hydrogen, and oxygen; their hydrocarbon backbones are used to make amino acids and other carbon-based molecules that can be assembled into larger molecules (such as proteins or DNA), used for example to form new cells. (HS-LS1-6) • As matter and energy flow through different organizational levels of living systems, chemical elements are recombined in different ways to form different products. (HS-LS1-6),(HS-LS1-7) • As a result of these chemical reactions, energy is transferred from one system of interacting molecules to another. Cellular respiration is a chemical process whereby the bonds of food molecules and oxygen molecules are broken and new compounds are formed that can transport energy to muscles. Cellular respiration also releases the energy needed to maintain body temperature despite ongoing energy transfer to the surrounding environment. (HS-LS1-7)
LS1.D: Information Processing	• Animals have body parts that capture and convey different kinds of information needed for growth and survival. Animals respond to these inputs with behaviors that help them survive. Plants also respond to some external inputs. (1-LS1-1)	• Different sense receptors are specialized for particular kinds of information, which may be then processed by the animal's brain. Animals are able to use their perceptions and memories to guide their actions. (4-LS1-2)	• Each sense receptor responds to different inputs (electromagnetic, mechanical, chemical), transmitting them as signals that travel along nerve cells to the brain. The signals are then processed in the brain, resulting in immediate behaviors or memories. (MS-LS1-8)	• N/A

N/A = Not applicable for this grade range

Disciplinary Core Ideas in Life Science (*continued*)

	Grades K–2	Grades 3–5	Grades 6–8	Grades 9–12
LS2: Ecosystems: Interactions, Energy, and Dynamics				
LS2.A: Interdependent Relationships in Ecosystems	• Plants depend on water and light to grow. (2-LS2-1) • Plants depend on animals for pollination or to move their seeds around. (2-LS2-2)	• The food of almost any kind of animal can be traced back to plants. Organisms are related in food webs in which some animals eat plants for food and other animals eat the animals that eat plants. Some organisms, such as fungi and bacteria, break down dead organisms (both plants or plant parts and animals) and therefore operate as "decomposers." Decomposition eventually restores (recycles) some materials back to the soil. Organisms can survive only in environments in which their particular needs are met. A healthy ecosystem is one in which multiple species of different types are each able to meet their needs in a relatively stable web of life. Newly introduced species can damage the balance of an ecosystem. (5-LS2-1)	• Organisms, and populations of organisms, are dependent on their environmental interactions both with other living things and with nonliving factors. (MS-LS2-1) • In any ecosystem, organisms and populations with similar requirements for food, water, oxygen, or other resources may compete with each other for limited resources, access to which consequently constrains their growth and reproduction. (MS-LS2-1) • Growth of organisms and population increases are limited by access to resources. (MS-LS2-1) • Similarly, predatory interactions may reduce the number of organisms or eliminate whole populations of organisms. Mutually beneficial interactions, in contrast, may become so interdependent that each organism requires the other for survival. Although the species involved in these competitive, predatory, and mutually beneficial interactions vary across ecosystems, the patterns of interactions of organisms with their environments, both living and nonliving, are shared. (MS-LS2-2)	• Ecosystems have carrying capacities, which are limits to the numbers of organisms and populations they can support. These limits result from such factors as the availability of living and nonliving resources and from challenges such as predation, competition, and disease. Organisms would have the capacity to produce populations of great size were it not for the fact that environments and resources are finite. This fundamental tension affects the abundance (number of individuals) of species in any given ecosystem. (HS-LS2-1),(HS-LS2-2)
LS2.B: Cycles of Matter and Energy Transfer in Ecosystems	• N/A	• Matter cycles between the air and soil and among plants, animals, and microbes as these organisms live and die. Organisms obtain gases and water from the environment and release waste matter (gas, liquid, or solid) back into the environment. (5-LS2-1)	• Food webs are models that demonstrate how matter and energy are transferred between producers, consumers, and decomposers as the three groups interact within an ecosystem. Transfers of matter into and out of the physical environment occur at every level. Decomposers recycle nutrients from dead plant or animal matter back to the soil in terrestrial environments or to the water in aquatic environments. The atoms that make up the organisms in an ecosystem are cycled repeatedly between the living and nonliving parts of the ecosystem. (MS-LS2-3)	• Photosynthesis and cellular respiration (including anaerobic processes) provide most of the energy for life processes. (HS-LS2-3) • Plants or algae form the lowest level of the food web. At each link upward in a food web, only a small fraction of the matter consumed at the lower level is transferred upward, to produce growth and release energy in cellular respiration at the higher level. Given this inefficiency, there are generally fewer organisms at higher levels of a food web. Some matter reacts to release energy for life functions, some matter is stored in newly made structures, and much is discarded. The chemical elements that make up the molecules of organisms pass through food webs and into and out of the atmosphere and soil, and they are combined and recombined in different ways. At each link in an ecosystem, matter and energy are conserved. (HS-LS2-4) • Photosynthesis and cellular respiration are important components of the carbon cycle, in which carbon is exchanged among the biosphere, atmosphere, oceans, and geosphere through chemical, physical, geologic, and biological processes. (HS-LS2-5)

N/A = Not applicable for this grade range

Disciplinary Core Ideas in Life Science (*continued*)

	Grades K–2	Grades 3–5	Grades 6–8	Grades 9–12
LS2.C: Ecosystem Dynamics, Functioning, and Resilience	• N/A	• When the environment changes in ways that affect a place's physical characteristics, temperature, or availability of resources, some organisms survive and reproduce, others move to new locations, yet others move into the transformed environment, and some die. (secondary to 3-LS4-4)	• Ecosystems are dynamic in nature; their characteristics can vary over time. Disruptions to any physical or biological component of an ecosystem can lead to shifts in all its populations. (MS-LS2-4) • Biodiversity describes the variety of species found in Earth's terrestrial and oceanic ecosystems. The completeness or integrity of an ecosystem's biodiversity is often used as a measure of its health. (MS-LS2-5)	• A complex set of interactions within an ecosystem can keep its numbers and types of organisms relatively constant over long periods of time under stable conditions. If a modest biological or physical disturbance to an ecosystem occurs, it may return to its more or less original status (i.e., the ecosystem is resilient), as opposed to becoming a very different ecosystem. Extreme fluctuations in conditions or the size of any population, however, can challenge the functioning of ecosystems in terms of resources and habitat availability. (HS-LS2-2),(HS-LS2-6) • Moreover, anthropogenic changes (induced by human activity) in the environment—including habitat destruction, pollution, introduction of invasive species, overexploitation, and climate change—can disrupt an ecosystem and threaten the survival of some species. (HS-LS2-7)
LS2.D: Social Interactions and Group Behavior	• N/A	• Being part of a group helps animals obtain food, defend themselves, and cope with changes. Groups may serve different functions and vary dramatically in size (Note: Moved from K–2.) (3-LS2-1)	• N/A	• Group behavior has evolved because membership can increase the chances of survival for individuals and their genetic relatives. (HS-LS2-8)
LS3: Heredity: Inheritance and Variation of Traits				
LS3.A: Inheritance of Traits	• Young animals are very much, but not exactly, like, their parents. Plants also are very much, but not exactly, like their parents. (1-LS3-1)	• Many characteristics of organisms are inherited from their parents. (3-LS3-1) • Other characteristics result from individuals' interactions with the environment, which can range from diet to learning. Many characteristics involve both inheritance and environment. (3-LS3-2)	• Genes are located in the chromosomes of cells, with each chromosome pair containing two variants of each of many distinct genes. Each distinct gene chiefly controls the production of specific proteins, which in turn affects the traits of the individual. Changes (mutations) to genes can result in changes to proteins, which can affect the structures and functions of the organism and thereby change traits. (MS-LS3-1) • Variations of inherited traits between parent and offspring arise from genetic differences that result from the subset of chromosomes (and therefore genes) inherited. (MS-LS3-2)	• Each chromosome consists of a single very long DNA molecule, and each gene on the chromosome is a particular segment of that DNA. The instructions for forming species' characteristics are carried in DNA. All cells in an organism have the same genetic content, but the genes used (expressed) by the cell may be regulated in different ways. Not all DNA codes for a protein; some segments of DNA are involved in regulatory or structural functions, and some have no as-yet known function. (HS-LS3-1)

N/A = Not applicable for this grade range

Disciplinary Core Ideas in Life Science (*continued*)

	Grades K–2	Grades 3–5	Grades 6–8	Grades 9–12
LS3.B: Variation of Traits	• Individuals of the same kind of plant or animal are recognizable as similar but can also vary in many ways. (1-LS3-1)	• Different organisms vary in how they look and function because they have different inherited information. (3-LS3-1) • The environment also affects the traits that an organism develops. (3-LS3-2)	• In sexually reproducing organisms, each parent contributes half of the genes acquired (at random) by the offspring. Individuals have two of each chromosome and hence two alleles of each gene, one acquired from each parent. These versions may be identical or may differ from each other. (MS-LS3-2) • In addition to variations that arise from sexual reproduction, genetic information can be altered because of mutations. Though rare, mutations may result in changes to the structure and function of proteins. Some changes are beneficial, others harmful, and some neutral to the organism. (MS-LS3-1)	• In sexual reproduction, chromosomes can sometimes swap sections during the process of meiosis (cell division), thereby creating new genetic combinations and thus more genetic variation. Although DNA replication is tightly regulated and remarkably accurate, errors do occur and result in mutations, which are also a source of genetic variation. Environmental factors can also cause mutations in genes, and viable mutations are inherited. (HS-LS3-2) • Environmental factors also affect expression of traits, and hence affect the probability of occurrences of traits in a population. Thus, the variation and distribution of traits observed depends on both genetic and environmental factors. (HS-LS3-2),(HS-LS3-3)
LS4: Biological Evolution: Unity and Diversity				
LS4.A: Evidence of Common Ancestry and Diversity	• N/A	• Some kinds of plants and animals that once lived on Earth are no longer found anywhere. (Note: Moved from K-2.) (3-LS4-1) • Fossils provide evidence about the types of organisms that lived long ago and also about the nature of their environments. (3-LS4-1)	• The collection of fossils and their placement in chronological order (e.g., through the location of the sedimentary layers in which they are found or through radioactive dating) is known as the fossil record. It documents the existence, diversity, extinction, and change of many life forms throughout the history of life on Earth. (MS-LS4-1) • Anatomical similarities and differences between various organisms living today and between them and organisms in the fossil record enable the reconstruction of evolutionary history and the inference of lines of evolutionary descent. (MS-LS4-2) • Comparison of the embryological development of different species also reveals similarities that show relationships not evident in the fully formed anatomy. (MS-LS4-3)	• Genetic information provides evidence of evolution. DNA sequences vary among species, but there are many overlaps; in fact, the ongoing branching that produces multiple lines of descent can be inferred by comparing the DNA sequences of different organisms. Such information is also derivable from the similarities and differences in amino acid sequences and from anatomical and embryological evidence. (HS-LS4-1)
LS4.B: Natural Selection	• N/A	• Sometimes the differences in characteristics between individuals of the same species provide advantages in surviving, finding mates, and reproducing. (3-LS4-2)	• Natural selection leads to the predominance of certain traits in a population, and the suppression of others. (MS-LS4-4) • In artificial selection, humans have the capacity to influence certain characteristics of organisms by selective breeding. One can choose desired parental traits determined by genes, which are then passed on to offspring. (MS-LS4-5)	• Natural selection occurs only if there is both (1) variation in the genetic information between organisms in a population and (2) variation in the expression of that genetic information—that is, trait variation—that leads to differences in performance among individuals. (HS-LS4-2),(HS-LS4-3) • The traits that positively affect survival are more likely to be reproduced, and thus are more common in the population. (HS-LS4-3)

N/A = Not applicable for this grade range

Disciplinary Core Ideas in Life Science (*continued*)

	Grades K–2	Grades 3–5	Grades 6–8	Grades 9–12
LS4.C: Adaptation	• N/A	• For any particular environment, some kinds of organisms survive well, some survive less well, and some cannot survive at all. (3-LS4-3)	• Adaptation by natural selection acting over generations is one important process by which species change over time in response to changes in environmental conditions. Traits that support successful survival and reproduction in the new environment become more common; those that do not become less common. Thus, the distribution of traits in a population changes. (MS-LS4-6)	• Evolution is a consequence of the interaction of four factors: (1) the potential for a species to increase in number, (2) the genetic variation of individuals in a species due to mutation and sexual reproduction, (3) competition for an environment's limited supply of the resources that individuals need in order to survive and reproduce, and (4) the ensuing proliferation of those organisms that are better able to survive and reproduce in that environment. (HS-LS4-2) • Natural selection leads to adaptation, that is, to a population dominated by organisms that are anatomically, behaviorally, and physiologically well suited to survive and reproduce in a specific environment. That is, the differential survival and reproduction of organisms in a population that have an advantageous heritable trait leads to an increase in the proportion of individuals in future generations that have the trait and to a decrease in the proportion of individuals that do not. (HS-LS4-3),(HS-LS4-4) • Adaptation also means that the distribution of traits in a population can change when conditions change. (HS-LS4-3) • Changes in the physical environment, whether naturally occurring or human induced, have thus contributed to the expansion of some species, the emergence of new distinct species as populations diverge under different conditions, and the decline–and sometimes the extinction–of some species. (HS-LS4-5),(HS-LS4-6) • Species become extinct because they can no longer survive and reproduce in their altered environment. If members cannot adjust to change that is too fast or drastic, the opportunity for the species' evolution is lost. (HS-LS4-5)

N/A = Not applicable for this grade range

Disciplinary Core Ideas in Life Science (*continued*)

	Grades K–2	Grades 3–5	Grades 6–8	Grades 9–12
LS4.D: Biodiversity and Humans	• There are many different kinds of living things in any area, and they exist in different places on land and in water. (2-LS4-1)	• Populations live in a variety of habitats, and change in those habitats affects the organisms living there. (3-LS4-4)	• Changes in biodiversity can influence humans' resources, such as food, energy, and medicines, as well as ecosystem services that humans rely on—for example, water purification and recycling. (secondary to MS-LS2-5)	• Biodiversity is increased by the formation of new species (speciation) and decreased by the loss of species (extinction). (secondary to HS-LS2-7) • Humans depend on the living world for the resources and other benefits provided by biodiversity. But human activity is also having adverse impacts on biodiversity through overpopulation, overexploitation, habitat destruction, pollution, introduction of invasive species, and climate change. Thus, sustaining biodiversity so that ecosystem functioning and productivity are maintained is essential to supporting and enhancing life on Earth. Sustaining biodiversity also aids humanity by preserving landscapes of recreational or inspirational value. (secondary to HS-LS2-7), (HS-LS4-6)

Disciplinary Core Ideas in Earth and Space Science

	Grades K–2	Grades 3–5	Grades 6–8	Grades 9–12
ESS1: Earth's Place in the Universe				
ESS1.A: The Universe and Its Stars	• Patterns of the motion of the Sun, Moon, and stars in the sky can be observed, described, and predicted. (1-ESS1-1)	• The Sun is a star that appears larger and brighter than other stars because it is closer. Stars range greatly in their distance from Earth. (5-ESS1-1)	• Patterns of the apparent motion of the Sun, the Moon, and stars in the sky can be observed, described, predicted, and explained with models. (MS-ESS1-1) • Earth and its solar system are part of the Milky Way galaxy, which is one of many galaxies in the universe. (MS-ESS1-2)	• The star called the Sun is changing and will burn out over a life span of approximately 10 billion years. (HS-ESS1-1) • The study of stars' light spectra and brightness is used to identify compositional elements of stars, their movements, and their distances from Earth. (HS-ESS1-2),(HS-ESS1-3) • The big bang theory is supported by observations of distant galaxies receding from our own, of the measured composition of stars and nonstellar gases, and of the maps of spectra of the primordial radiation (cosmic microwave background) that still fills the universe. (HS-ESS1-2) • Other than the hydrogen and helium formed at the time of the big bang, nuclear fusion within stars produces all atomic nuclei lighter than and including iron, and the process releases electromagnetic energy. Heavier elements are produced when certain massive stars achieve a supernova stage and explode. (HS-ESS1- 2),(HS-ESS1-3)
ESS1.B: Earth and the Solar System	• Seasonal patterns of sunrise and sunset can be observed, described, and predicted. (1-ESS1-2)	• The orbits of Earth around the Sun and of the Moon around Earth, together with the rotation of Earth about an axis between its North and South poles, cause observable patterns. These include day and night; daily changes in the length and direction of shadows; and different positions of the Sun, Moon, and stars at different times of the day, month, and year. (5-ESS1-2)	• The solar system consists of the Sun and a collection of objects, including planets, their moons, and asteroids that are held in orbit around the Sun by its gravitational pull on them. (MS-ESS1-2),(MS-ESS1-3) • This model of the solar system can explain eclipses of the Sun and the Moon. Earth's spin axis is fixed in direction over the short term but tilted relative to its orbit around the Sun. The seasons are a result of that tilt and are caused by the differential intensity of sunlight on different areas of Earth across the year. (MS-ESS1-1) • The solar system appears to have formed from a disk of dust and gas, drawn together by gravity. (MS-ESS1-2)	• Kepler's laws describe common features of the motions of orbiting objects, including their elliptical paths around the Sun. Orbits may change due to the gravitational effects from, or collisions with, other objects in the solar system. (HS-ESS1-4) • Cyclical changes in the shape of Earth's orbit around the Sun, together with changes in the tilt of the planet's axis of rotation, both occurring over hundreds of thousands of years, have altered the intensity and distribution of sunlight falling on the Earth. These phenomena cause a cycle of ice ages and other gradual climate changes. (secondary to HS-ESS2-4)

Disciplinary Core Ideas in Earth and Space Science (*continued*)

	Grades K–2	Grades 3–5	Grades 6–8	Grades 9–12
ESS1.C: The History of Planet Earth	• Some events happen very quickly; others occur very slowly, over a time period much longer than one can observe. (2-ESS1-1)	• Local, regional, and global patterns of rock formations reveal changes over time due to Earth forces, such as earthquakes. The presence and location of certain fossil types indicate the order in which rock layers were formed. (4-ESS1-1)	• The geologic time scale interpreted from rock strata provides a way to organize Earth's history. Analyses of rock strata and the fossil record provide only relative dates, not an absolute scale. (MS-ESS1-4) • Tectonic processes continually generate new ocean seafloor at ridges and destroy old seafloor at trenches. (HS.ESS1.C GBE), (secondary to MS-ESS2-3)	• Continental rocks, which can be older than 4 billion years, are generally much older than the rocks of the ocean floor, which are less than 200 million years old. (HS-ESS1-5) • Although active geologic processes, such as plate tectonics and erosion, have destroyed or altered most of the very early rock record on Earth, other objects in the solar system, such as lunar rocks, asteroids, and meteorites, have changed little over billions of years. Studying these objects can provide information about Earth's formation and early history. (HS-ESS1-6)
ESS2: Earth's Systems				
ESS2.A: Earth Materials and Systems	• Wind and water can change the shape of the land. (2-ESS2-1)	• Earth's major systems are the geosphere (solid and molten rock, soil, and sediments), the hydrosphere (water and ice), the atmosphere (air), and the biosphere (living things, including humans). These systems interact in multiple ways to affect Earth's surface materials and processes. The ocean supports a variety of ecosystems and organisms, shapes landforms, and influences climate. Winds and clouds in the atmosphere interact with the landforms to determine patterns of weather. (5-ESS2-1) • Rainfall helps to shape the land and affects the types of living things found in a region. Water, ice, wind, living organisms, and gravity break rocks, soils, and sediments into smaller particles and move them around. (4-ESS2-1)	• All Earth processes are the result of energy flowing and matter cycling within and among the planet's systems. This energy is derived from the Sun and Earth's hot interior. The energy that flows and matter that cycles produce chemical and physical changes in Earth's materials and living organisms. (MS-ESS2-1) • The planet's systems interact over scales that range from microscopic to global in size, and they operate over fractions of a second to billions of years. These interactions have shaped Earth's history and will determine its future. (MS-ESS2-2)	• Earth's systems, being dynamic and interacting, cause feedback effects that can increase or decrease the original changes. (HS-ESS2-1),(HS-ESS2-2) • Evidence from deep probes and seismic waves, reconstructions of historical changes in Earth's surface and its magnetic field, and an understanding of physical and chemical processes lead to a model of Earth with a hot but solid inner core, a liquid outer core, and a solid mantle and crust. Motions of the mantle and its plates occur primarily through thermal convection, which involves the cycling of matter due to the outward flow of energy from Earth's interior and gravitational movement of denser materials toward the interior. (HS-ESS2-3) • The geologic record shows that changes to global and regional climate can be caused by interactions among changes in the Sun's energy output or Earth's orbit, tectonic events, ocean circulation, volcanic activity, glaciers, vegetation, and human activities. These changes can occur on a variety of time scales from sudden (e.g., volcanic ash clouds) to intermediate (ice ages) to very long-term tectonic cycles. (HS-ESS2-4)

Disciplinary Core Ideas in Earth and Space Science (*continued*)

	Grades K–2	Grades 3–5	Grades 6–8	Grades 9–12
ESS2.B: Plate Tectonics and Large-Scale System Interactions	• Maps show where things are located. One can map the shapes and kinds of land and water in any area. (2-ESS2-2)	• The locations of mountain ranges, deep ocean trenches, ocean floor structures, earthquakes, and volcanoes occur in patterns. Most earthquakes and volcanoes occur in bands that are often along the boundaries between continents and oceans. Major mountain chains form inside continents or near their edges. Maps can help locate the different land and water features of Earth. (4-ESS2-2)	• Maps of ancient land and water patterns, based on investigations of rocks and fossils, make clear how Earth's plates have moved great distances, collided, and spread apart. (MS-ESS2-3)	• Plate tectonics is the unifying theory that explains the past and current movements of the rocks at Earth's surface and provides a framework for understanding its geologic history. (ESS2.B Grade 8 GBE), (HS-ESS2-1), (secondary to HS-ESS1-5) • Plate movements are responsible for most continental and ocean-floor features and for the distribution of most rocks and minerals within Earth's crust. (ESS2.B Grade 8 GBE), (HS-ESS2-1) • The radioactive decay of unstable isotopes continually generates new energy within Earth's crust and mantle, providing the primary source of the heat that drives mantle convection. Plate tectonics can be viewed as the surface expression of mantle convection. (HS-ESS2-3)
ESS2.C: The Roles of Water in Earth's Surface Processes	• Water is found in the ocean, rivers, lakes, and ponds. Water exists as solid ice and in liquid form. (2-ESS2-3)	• Nearly all of Earth's available water is in the ocean. Most fresh water is in glaciers or underground; only a tiny fraction is in streams, lakes, wetlands, and the atmosphere. (5-ESS2-2)	• Water continually cycles among land, ocean, and atmosphere via transpiration, evaporation, condensation and crystallization, and precipitation, as well as downhill flows on land. (MS-ESS2-4) • The complex patterns of the changes and the movement of water in the atmosphere, determined by winds, landforms, and ocean temperatures and currents, are major determinants of local weather patterns. (MS-ESS2-5) • Global movements of water and its changes in form are propelled by sunlight and gravity. (MS-ESS2-4) • Variations in density due to variations in temperature and salinity drive a global pattern of interconnected ocean currents. (MS-ESS2-6) • Water's movements—both on the land and underground—cause weathering and erosion, which change the land's surface features and create underground formations. (MS-ESS2-2)	• The abundance of liquid water on Earth's surface and its unique combination of physical and chemical properties are central to the planet's dynamics. These properties include water's exceptional capacity to absorb, store, and release large amounts of energy, transmit sunlight, expand upon freezing, dissolve and transport materials, and lower the viscosities and melting points of rocks. (HS-ESS2-5)

Disciplinary Core Ideas in Earth and Space Science *(continued)*

	Grades K–2	Grades 3–5	Grades 6–8	Grades 9–12
ESS2.D: Weather and Climate	• Weather is the combination of sunlight, wind, snow or rain, and temperature in a particular region at a particular time. People measure these conditions to describe and record the weather and to notice patterns over time. (K-ESS2-1)	• Scientists record patterns of the weather across different times and areas so that they can make predictions about what kind of weather might happen next. (3-ESS2-1) • Climate describes a range of an area's typical weather conditions and the extent to which those conditions vary over years. (3-ESS2-2)	• Weather and climate are influenced by interactions involving sunlight, the ocean, the atmosphere, ice, landforms, and living things. These interactions vary with latitude, altitude, and local and regional geography, all of which can affect oceanic and atmospheric flow patterns. (MS-ESS2-6) • Because these patterns are so complex, weather can only be predicted probabilistically. (MS-ESS2-5) • The ocean exerts a major influence on weather and climate by absorbing energy from the Sun, releasing it over time, and globally redistributing it through ocean currents. (MS-ESS2-6)	• The foundation for Earth's global climate systems is the electromagnetic radiation from the Sun, as well as its reflection, absorption, storage, and redistribution among the atmosphere, ocean, and land systems, and this energy's re-radiation into space. (HS-ESS2-4) • Gradual atmospheric changes were due to plants and other organisms that captured carbon dioxide and released oxygen. (HS-ESS2-6),(HS-ESS2-7) • Changes in the atmosphere due to human activity have increased carbon dioxide concentrations and thus affect climate. (HS-ESS2-6),(HS-ESS2-4) • Current models predict that, although future regional climate changes will be complex and varied, average global temperatures will continue to rise. The outcomes predicted by global climate models strongly depend on the amounts of human-generated greenhouse gases added to the atmosphere each year and by the ways in which these gases are absorbed by the ocean and biosphere. (secondary to HS-ESS3-6)
ESS2.E: Biogeology	• Plants and animals can change their environment. (K-ESS2-2)	• Living things affect the physical characteristics of their regions. (4-ESS2-1)	• N/A	• The many dynamic and delicate feedbacks between the biosphere and other Earth systems cause a continual co-evolution of Earth's surface and the life that exists on it. (HS-ESS2-7)
ESS3: Earth and Human Activity				
ESS3.A: Natural Resources	• Living things need water, air, and resources from the land, and they live in places that have the things they need. Humans use natural resources for everything they do. (K-ESS3-1)	• Energy and fuels that humans use are derived from natural sources, and their use affects the environment in multiple ways. Some resources are renewable over time, and others are not. (4-ESS3-1)	• Humans depend on Earth's land, ocean, atmosphere, and biosphere for many different resources. Minerals, fresh water, and biosphere resources are limited, and many are not renewable or replaceable over human lifetimes. These resources are distributed unevenly around the planet as a result of past geologic processes. (MS-ESS3-1)	• Resource availability has guided the development of human society. (HS-ESS3-1) • All forms of energy production and other resource extraction have associated economic, social, environmental, and geopolitical costs and risks as well as benefits. New technologies and social regulations can change the balance of these factors. (HS-ESS3-2)

N/A = Not applicable for this grade range

Disciplinary Core Ideas in Earth and Space Science (*continued*)

	Grades K–2	Grades 3–5	Grades 6–8	Grades 9–12
ESS3.B: Natural Hazards	• Some kinds of severe weather are more likely than others in a given region. Weather scientists forecast severe weather so that the communities can prepare for and respond to these events. (K-ESS3-2)	• A variety of natural hazards result from natural processes. Humans cannot eliminate natural hazards but can take steps to reduce their impacts. (3-ESS3-1), (4-ESS3-2)	• Mapping the history of natural hazards in a region, combined with an understanding of related geologic forces, can help forecast the locations and likelihoods of future events. (MS-ESS3-2)	• Natural hazards and other geologic events have shaped the course of human history; they have significantly altered the sizes of human populations and have driven human migrations. (HS-ESS3-1)
ESS3.C: Human Impacts on Earth Systems	• Things that people do to live comfortably can affect the world around them. But they can make choices that reduce their impacts on the land, water, air, and other living things. (K-ESS3-3), (secondary to K-ESS2-2)	• Human activities in agriculture, industry, and everyday life have had major effects on the land, vegetation, streams, ocean, air, and even outer space. But individuals and communities are doing things to help protect Earth's resources and environments. (5-ESS3-1)	• Human activities have significantly altered the biosphere, sometimes damaging or destroying natural habitats and causing the extinction of other species. But changes to Earth's environments can have different impacts (negative and positive) for different living things. (MS-ESS3-3) • Typically as human populations and per capita consumption of natural resources increase, so do the negative impacts on Earth unless the activities and technologies involved are engineered otherwise. (MS-ESS3-3),(MS-ESS3-4)	• The sustainability of human societies and the biodiversity that supports them requires responsible management of natural resources. (HS-ESS3-3) • Scientists and engineers can make major contributions by developing technologies that produce less pollution and waste and that preclude ecosystem degradation. (HS-ESS3-4)
ESS3.D: Global Climate Change	• N/A	• N/A	• Human activities, such as the release of greenhouse gases from burning fossil fuels, are major factors in the current rise in Earth's mean surface temperature (global warming). Reducing the level of climate change and reducing human vulnerability to whatever climate changes do occur depend on the understanding of climate science, engineering capabilities, and other kinds of knowledge, such as understanding of human behavior, and on applying that knowledge wisely in decisions and activities. (MS-ESS3-5)	• Though the magnitudes of human impacts are greater than they have ever been, so too are human abilities to model, predict, and manage current and future impacts. (HS-ESS3-5) • Through computer simulations and other studies, important discoveries are still being made about how the ocean, the atmosphere, and the biosphere interact and are modified in response to human activities. (HS-ESS3-6)

N/A = Not applicable for this grade range

Disciplinary Core Ideas in Engineering Design

	Grades K–2	Grades 3–5	Grades 6–8	Grades 9–12
ETS1: Engineering Design				
ETS1.A: Defining and Delimiting Engineering Problems	• A situation that people want to change or create can be approached as a problem to be solved through engineering. Such problems may have many acceptable solutions. (K–2-ETS1-1) (secondary to K-PS2-2) • Asking questions, making observations, and gathering information are helpful in thinking about problems. (K–2-ETS1-1), (secondary to K-ESS3-2) • Before beginning to design a solution, it is important to clearly understand the problem. (K–2-ETS1-1)	• Possible solutions to a problem are limited by available materials and resources (constraints). The success of a designed solution is determined by considering the desired features of a solution (criteria). Different proposals for solutions can be compared on the basis of how well each one meets the specified criteria for success or how well each takes the constraints into account. (3–5-ETS1-1), (secondary to 4-PS3-4)	• The more precisely a design task's criteria and constraints can be defined, the more likely it is that the designed solution will be successful. Specification of constraints includes consideration of scientific principles and other relevant knowledge that is likely to limit possible solutions. (MS-ETS1-1), (secondary to MS-PS3-3)	• Criteria and constraints also include satisfying any requirements set by society, such as taking issues of risk mitigation into account, and they should be quantified to the extent possible and stated in such a way that one can tell if a given design meets them. (HS-ETS1-1), (secondary to HS-PS2-3), (secondary to HS-PS3-3) • Humanity faces major global challenges today, such as the need for supplies of clean water and food or for energy sources that minimize pollution, which can be addressed through engineering. These global challenges also may have manifestations in local communities. (HS-ETS1-1)

Disciplinary Core Ideas in Engineering Design (*continued*)

	Grades K–2	Grades 3–5	Grades 6–8	Grades 9–12
ETS1.B: Developing Possible Solutions	• Designs can be conveyed through sketches, drawings, or physical models. These representations are useful in communicating ideas for a problem's solutions to other people. (K–2-ETS1-2), (secondary to K-ESS3-3), (secondary to 2-LS2-2)	• Research on a problem should be carried out before beginning to design a solution. Testing a solution involves investigating how well it performs under a range of likely conditions. (3–5-ETS1-2) (secondary to 4-ESS3-2) • Tests are often designed to identify failure points or difficulties, which suggest the elements of the design that need to be improved. (3–5-ETS1-3) • At whatever stage, communicating with peers about proposed solutions is an important part of the design process, and shared ideas can lead to improved designs. (3–5-ETS1-2)	• A solution needs to be tested, and then modified on the basis of the test results, in order to improve it. (MS-ETS1-4), (secondary to MS-PS1-6) • There are systematic processes for evaluating solutions with respect to how well they meet criteria and constraints of a problem. MS-ETS1-2), (MS-ETS1-3), (secondary to MS-PS3-3), (secondary to MS-LS2-5) • Sometimes parts of different solutions can be combined to create a solution that is better than any of its predecessors. (MS-ETS1-3) • Models of all kinds are important for testing solutions. (MS-ETS1-4)	• When evaluating solutions it is important to take into account a range of constraints including cost, safety, reliability, and aesthetics and to consider social, cultural, and environmental impacts. (secondary to HS-LS2-7), (secondary to HS-LS4-6), (secondary to HS-ESS3-2), (secondary to HS-ESS3-4), (HS-ETS1-3) • Both physical models and computers can be used in various ways to aid in the engineering design process. Computers are useful for a variety of purposes, such as running simulations to test different ways of solving a problem or to see which one is most efficient or economical; and in making a persuasive presentation to a client about how a given design will meet his or her needs. (HS-ETS1-4), (secondary to HS-LS4-6)
ETS1.C: Optimizing the Design Solution	• Because there is always more than one possible solution to a problem, it is useful to compare and test designs. (K–2-ETS1-3), (secondary to 2-ESS2-1)	• Different solutions need to be tested in order to determine which of them best solves the problem, given the criteria and the constraints. (3–5-ETS1-3), (secondary to 4-PS4-3)	• Although one design may not perform the best across all tests, identifying the characteristics of the design that performed the best in each test can provide useful information for the redesign process—that is, some of the characteristics may be incorporated into the new design. (MS-ETS1-3), (secondary to MS-PS1-6) • The iterative process of testing the most promising solutions and modifying what is proposed on the basis of the test results leads to greater refinement and ultimately to an optimal solution. (MS-ETS1-4), (secondary to MS-PS1-6)	• Criteria may need to be broken down into simpler ones that can be approached systematically, and decisions about the priority of certain criteria over others (trade-offs) may be needed. (HS-ETS1-2), (secondary to HS-PS1-6), (secondary to HS-PS2-3)

Connections to the Nature of Science

Understandings About the Nature of Science Most Closely Associated With Practices				
Category	**Grades K–2**	**Grades 3–5**	**Grades 6–8**	**Grades 9–12**
Scientific Investigations Use a Variety of Methods	• Science investigations begin with a question. • Science uses different ways to study the world.	• Science methods are determined by questions. • Science investigations use a variety of methods, tools, and techniques.	• Science investigations use a variety of methods and tools to make measurements and observations. • Science investigations are guided by a set of values to ensure accuracy of measurements, observations, and objectivity of findings. • Science depends on evaluating proposed explanations. • Scientific values function as criteria in distinguishing between science and non-science.	• Science investigations use diverse methods and do not always use the same set of procedures to obtain data. • New technologies advance scientific knowledge. • Scientific inquiry is characterized by a common set of values that include logical thinking, precision, open-mindedness, objectivity, skepticism, replicability of results, and honest and ethical reporting of findings. • The discourse practices of science are organized around disciplinary domains that share exemplars for making decisions regarding the values, instruments, methods, models, and evidence to adopt and use. • Scientific investigations use a variety of methods, tools, and techniques to revise and produce new knowledge.
Scientific Knowledge Is Based on Empirical Evidence	• Scientists look for patterns and order when making observations about the world.	• Science findings are based on recognizing patterns. • Science uses tools and technologies to make accurate measurements and observations.	• Science knowledge is based upon logical and conceptual connections between evidence and explanations. • Science disciplines share common rules of obtaining and evaluating empirical evidence.	• Science knowledge is based on empirical evidence. • Science disciplines share common rules of evidence used to evaluate explanations about natural systems. • Science includes the process of coordinating patterns of evidence with current theory. • Science arguments are strengthened by multiple lines of evidence supporting a single explanation.

Connections to the Nature of Science (*continued*)

Category	Grades K–2	Grades 3–5	Grades 6–8	Grades 9–12
Scientific Knowledge Is Open to Revision in Light of New Evidence	• Science knowledge can change when new information is found.	• Science explanations can change based on new evidence.	• Scientific explanations are subject to revision and improvement in light of new evidence. • The certainty and durability of science findings vary. • Science findings are frequently revised and/or reinterpreted based on new evidence.	• Scientific explanations can be probabilistic. • Most scientific knowledge is quite durable but is, in principle, subject to change based on new evidence and/or reinterpretation of existing evidence. • Scientific argumentation is a mode of logical discourse used to clarify the strength of relationships between ideas and evidence that may result in revision of an explanation.
Science Models, Laws, Mechanisms, and Theories Explain Natural Phenomena	• Science uses drawings, sketches, and models as a way to communicate ideas. • Science searches for cause-and effect relationships to explain natural events.	• Science theories are based on a body of evidence and many tests. • Science explanations describe the mechanisms for natural events.	• Theories are explanations for observable phenomena. • Science theories are based on a body of evidence developed over time. • Laws are regularities or mathematical descriptions of natural phenomena. • A hypothesis is used by scientists as an idea that may contribute important new knowledge for the evaluation of a scientific theory. • The term “theory” as used in science is very different from the common use outside of science.	• Theories and laws provide explanations in science, but theories do not with time become laws or facts. • A scientific theory is a substantiated explanation of some aspect of the natural world, based on a body of facts that have been repeatedly confirmed through observation and experiment, and the science community validates each theory before it is accepted. If new evidence is discovered that the theory does not accommodate, the theory is generally modified in light of this new evidence. • Models, mechanisms, and explanations collectively serve as tools in the development of a scientific theory. • Laws are statements or descriptions of the relationships among observable phenomena. • Scientists often use hypotheses to develop and test theories and explanations.

Connections to the Nature of Science (*continued*)

Understandings About the Nature of Science Most Closely Associated With Crosscutting Concepts				
Category	**Grades K–2**	**Grades 3–5**	**Grades 6–8**	**Grades 9–12**
Science Is a Way of Knowing	• Science knowledge helps us know about the world.	• Science is both a body of knowledge and processes that add new knowledge. • Science is a way of knowing that is used by many people.	• Science is both a body of knowledge and the processes and practices used to add to that body of knowledge. • Science knowledge is cumulative and many people, from many generations and nations, have contributed to science knowledge. • Science is a way of knowing used by many people, not just scientists.	• Science is both a body of knowledge that represents a current understanding of natural systems and the processes used to refine, elaborate, revise, and extend this knowledge. • Science is a unique way of knowing, and there are other ways of knowing. • Science distinguishes itself from other ways of knowing through use of empirical standards, logical arguments, and skeptical review. • Science knowledge has a history that includes the refinement of, and changes to, theories, ideas, and beliefs over time.
Scientific Knowledge Assumes an Order and Consistency in Natural Systems	• Science assumes natural events happen today as they happened in the past. • Many events are repeated.	• Science assumes consistent patterns in natural systems. • Basic laws of nature are the same everywhere in the universe.	• Science assumes that objects and events in natural systems occur in consistent patterns that are understandable through measurement and observation. • Science carefully considers and evaluates anomalies in data and evidence.	• Scientific knowledge is based on the assumption that natural laws operate today as they did in the past and they will continue to do so in the future. • Science assumes the universe is a vast single system in which basic laws are consistent.
Science Is a Human Endeavor	• People have practiced science for a long time. • Men and women of diverse backgrounds are scientists and engineers.	• Men and women from all cultures and backgrounds choose careers as scientists and engineers. • Most scientists and engineers work in teams. • Science affects everyday life. • Creativity and imagination are important to science.	• Men and women from different social, cultural, and ethnic backgrounds work as scientists and engineers. • Scientists and engineers rely on human qualities such as persistence, precision, reasoning, logic, imagination, and creativity. • Scientists and engineers are guided by habits of mind such as intellectual honesty, tolerance of ambiguity, skepticism, and openness to new ideas. • Advances in technology influence the progress of science, and science has influenced advances in technology.	• Scientific knowledge is a result of human endeavor, imagination, and creativity. • Individuals and teams from many nations and cultures have contributed to science and to advances in engineering. • Scientists' backgrounds, theoretical commitments, and fields of endeavor influence the nature of their findings. • Technological advances have influenced the progress of science, and science has influenced advances in technology. • Science and engineering are influenced by society, and society is influenced by science and engineering.
Science Addresses Questions About the Natural and Material World	• Scientists study the natural and material world.	• Science findings are limited to questions that can be answered with empirical evidence.	• Scientific knowledge is constrained by human capacity, technology, and materials. • Science limits its explanations to systems that lend themselves to observation and empirical evidence. • Science knowledge can describe consequences of actions but is not responsible for society's decisions.	• Not all questions can be answered by science. • Science and technology may raise ethical issues for which science, by itself, does not provide answers and solutions. • Science knowledge indicates what can happen in natural systems—not what should happen. The latter involves ethics, values, and human decisions about the use of knowledge. • Many decisions are not made using science alone, but rely on social and cultural contexts to resolve issues.

Connections to Engineering, Technology, and Applications of Science

Grades K–2	Grades 3–5	Grades 6–8	Grades 9–12
Interdependence of Science, Engineering, and Technology			
• Science and engineering involve the use of tools to observe and measure things.	• Science and technology support each other. • Tools and instruments are used to answer scientific questions, while scientific discoveries lead to the development of new technologies.	• Engineering advances have led to important discoveries in virtually every field of science, and scientific discoveries have led to the development of entire industries and engineered systems. • Science and technology drive each other forward.	• Science and engineering complement each other in the cycle known as research and development (R&D). • Many R&D projects may involve scientists, engineers, and others with wide ranges of expertise.
Influence of Science, Engineering, and Technology on Society and the Natural World			
• Every human-made product is designed by applying some knowledge of the natural world and is built by using natural materials. • Taking natural materials to make things impacts the environment.	• People's needs and wants change over time, as do their demands for new and improved technologies. • Engineers improve existing technologies or develop new ones to increase their benefits, decrease known risks, and meet societal demands. • When new technologies become available, they can bring about changes in the way people live and interact with one another.	• All human activity draws on natural resources and has both short- and long-term consequences, positive as well as negative, for the health of people and the natural environment. • The uses of technologies and any limitations on their use are driven by individual or societal needs, desires, and values; by the findings of scientific research; and by differences in such factors as climate, natural resources, and economic conditions. • Technology use varies over time and from region to region.	• Modern civilization depends on major technological systems, such as agriculture, health, water, energy, transportation, manufacturing, construction, and communications. • Engineers continuously modify these systems to increase benefits while decreasing costs and risks. • New technologies can have deep impacts on society and the environment, including some that were not anticipated. • Analysis of costs and benefits is a critical aspect of decisions about technology.

CHAPTER 3
Focus on Grades K–2

Science and Engineering Practices

Asking Questions and Defining Problems for Grades K–2

Asking questions and defining problems in K–2 builds on prior experiences and progresses to simple descriptive questions that can be tested.

- Ask questions based on observations to find more information about the natural and/or designed world(s).
- Ask and/or identify questions that can be answered by an investigation.
- Define a simple problem that can be solved through the development of a new or improved object or tool.

Developing and Using Models for Grades K–2

Modeling in K–2 builds on prior experiences and progresses to include using and developing models (i.e., diagram, drawing, physical replica, diorama, dramatization, or storyboard) that represent concrete events or design solutions.

- Distinguish between a model and the actual object, process, and/or events the model represents.
- Compare models to identify common features and differences.
- Develop and/or use a model to represent amounts, relationships, relative scales (bigger, smaller), and/or patterns in the natural and designed world(s).
- Develop a simple model based on evidence to represent a proposed object or tool.

Planning and Carrying Out Investigations for Grades K–2

Planning and carrying out investigations to answer questions or test solutions to problems in K–2 builds on prior experiences and progresses to simple investigations, based on fair tests, which provide data to support explanations or design solutions.

- With guidance, plan and conduct an investigation in collaboration with peers (for K).
- Plan and conduct an investigation collaboratively to produce data to serve as the basis for evidence to answer a question.
- Evaluate different ways of observing and/or measuring a phenomenon to determine which way can answer a question.
- Make observations (firsthand or from media) and/or measurements to collect data that can be used to make comparisons.
- Make observations (firsthand or from media) and/or measurements of a proposed object or tool or solution to determine if it solves a problem or meets a goal.
- Make predictions based on prior experiences.

Analyzing and Interpreting Data for Grades K–2

Analyzing data in K–2 builds on prior experiences and progresses to collecting, recording, and sharing observations.

- Record information (observations, thoughts, and ideas).
- Use and share pictures, drawings, and/or writings of observations.
- Use observations (firsthand or from media) to describe patterns and/or relationships in the natural and designed world(s) in order to answer scientific questions and solve problems.
- Compare predictions (based on prior experiences) to what occurred (observable events).
- Analyze data from tests of an object or tool to determine if it works as intended.

Using Mathematics and Computational Thinking for Grades K–2

Mathematical and computational thinking in K–2 builds on prior experience and progresses to recognizing that mathematics can be used to describe the natural and designed world(s).

- Use counting and numbers to identify and describe patterns in the natural and designed world(s).
- Describe, measure, and/or compare quantitative attributes of different objects and display the data using simple graphs.
- Use quantitative data to compare two alternative solutions to a problem.

Science and Engineering Practices (*continued*)

Constructing Explanations and Designing Solutions for Grades K–2
Constructing explanations and designing solutions in K–2 builds on prior experiences and progresses to the use of evidence and ideas in constructing evidence-based accounts of natural phenomena and designing solutions. • Use information from observations (firsthand and from media) to construct an evidence-based account for natural phenomena. • Use tools and/or materials to design and/or build a device that solves a specific problem or a solution to a specific problem. • Generate and/or compare multiple solutions to a problem.
Engaging in Argument From Evidence for Grades K–2
Engaging in argument from evidence in K–2 builds on prior experiences and progresses to comparing ideas and representations about the natural and designed world(s). • Identify arguments that are supported by evidence. • Distinguish between explanations that account for all gathered evidence and those that do not. • Analyze why some evidence is relevant to a scientific question and some is not. • Distinguish between opinions and evidence in one's own explanations. • Listen actively to arguments to indicate agreement or disagreement based on evidence, and/or to retell the main points of the argument. • Construct an argument with evidence to support a claim. • Make a claim about the effectiveness of an object, tool, or solution that is supported by relevant evidence.
Obtaining, Evaluating, and Communicating Information for Grades K–2
Obtaining, evaluating, and communicating information in K–2 builds on prior experiences and uses observations and texts to communicate new information. • Read grade-appropriate texts and/or use media to obtain scientific and/or technical information to determine patterns in and/or evidence about the natural and designed world(s). • Describe how specific images (e.g., a diagram showing how a machine works) support a scientific or engineering idea. • Obtain information using various texts, text features (e.g., headings, tables of contents, glossaries, electronic menus, icons), and other media that will be useful in answering a scientific question and/or supporting a scientific claim. • Communicate information or design ideas and/or solutions with others in oral and/or written forms using models, drawings, writing, or numbers that provide detail about scientific ideas, practices, and/or design ideas.

Crosscutting Concepts and Connections to Engineering, Technology, and Applications of Science

Crosscutting Concepts for Grades K–2	
Patterns	• Patterns in the natural and human designed world can be observed, used to describe phenomena, and used as evidence.
Cause and Effect: Mechanism and Prediction	• Events have causes that generate observable patterns. • Simple tests can be designed to gather evidence to support or refute student ideas about causes.
Scale, Proportion, and Quantity	• Relative scales allow objects and events to be compared and described (e.g., bigger and smaller, hotter and colder, faster and slower). • Standard units are used to measure length.
Systems and System Models	• Objects and organisms can be described in terms of their parts. • Systems in the natural and designed world have parts that work together.
Energy and Matter: Flows, Cycles, and Conservation	• Objects may break into smaller pieces, be put together into larger pieces, or change shapes.
Structure and Function	• The shape and stability of structures of natural and designed objects are related to their function(s).
Stability and Change	• Some things stay the same while other things change. • Things may change slowly or rapidly.
Connections to Engineering, Technology, and Applications of Science for Grades K–2	
Interdependence of Science, Engineering, and Technology	• Science and engineering involve the use of tools to observe and measure things.
Influence of Science, Engineering, and Technology on Society and the Natural World	• Every human-made product is designed by applying some knowledge of the natural world and is built by using natural materials. • Taking natural materials to make things impacts the environment.

Connections to the Nature of Science

Understandings Most Closely Associated With Practices for Grades K–2	
Scientific Investigations Use a Variety of Methods	• Science investigations begin with a question. • Science uses different ways to study the world.
Scientific Knowledge Is Based on Empirical Evidence	• Scientists look for patterns and order when making observations about the world.
Scientific Knowledge Is Open to Revision in Light of New Evidence	• Science knowledge can change when new information is found.
Science Models, Laws, Mechanisms, and Theories Explain Natural Phenomena	• Science uses drawings, sketches, and models as a way to communicate ideas. • Science searches for cause-and-effect relationships to explain natural events.
Understandings Most Closely Associated With Crosscutting Concepts for Grades K–2	
Science Is a Way of Knowing	• Science knowledge helps us know about the world.
Scientific Knowledge Assumes an Order and Consistency in Natural Systems	• Science assumes natural events happen today as they happened in the past. • Many events are repeated.
Science Is a Human Endeavor	• People have practiced science for a long time. • Men and women of diverse backgrounds are scientists and engineers.
Science Addresses Questions About the Natural and Material World	• Scientists study the natural and material world.

Performance Expectations and Disciplinary Core Ideas for Kindergarten

Performance Expectations (PEs)	Disciplinary Core Ideas (DCIs)
K-LS1-1. Use observations to describe patterns of what plants and animals (including humans) need to survive. **Clarification Statement:** Examples of patterns could include that animals need to take in food but plants do not, the different kinds of food needed by different types of animals, the requirement of plants to have light, and that all living things need water.	**LS1.C. Organization for Matter and Energy Flow in Organisms** All animals need food in order to live and grow. They obtain their food from plants or from other animals. Plants need water and light to live and grow.
K-ESS2-1. Use and share observations of local weather conditions to describe patterns over time. **Clarification Statement:** Examples of qualitative observations could include descriptions of the weather (such as sunny, cloudy, rainy, and warm); examples of quantitative observations could include numbers of sunny, windy, and rainy days in a month. Examples of patterns could include that it is usually cooler in the morning than in the afternoon and the number of sunny days versus cloudy days in different months. **Assessment Boundary:** Assessment of quantitative observations limited to whole numbers and relative measures such as warmer/cooler.	**ESS2.D. Weather and Climate** Weather is the combination of sunlight, wind, snow or rain, and temperature in a particular region at a particular time. People measure these conditions to describe and record the weather and to notice patterns over time.
K-ESS2-2. Construct an argument supported by evidence for how plants and animals (including humans) can change the environment to meet their needs. **Clarification Statement:** Examples of plants and animals changing their environment could include a squirrel digging in the ground to hide its food and tree roots breaking concrete.	**ESS2.E. Biogeology** Plants and animals can change their environment.
K-ESS3-1. Use a model to represent the relationship between the needs of different plants and animals (including humans) and the places they live. **Clarification Statement:** Examples of relationships could include that deer eat buds and leaves so they usually live in forested areas and that grasses need sunlight so they often grow in meadows. Plants, animals, and their surroundings make up a system.	**ESS3.A. Natural Resources** Living things need water, air, and resources from the land, and they live in places that have the things they need. Humans use natural resources for everything they do.
K-ESS3-2. Ask questions to obtain information about the purpose of weather forecasting to prepare for, and respond to, severe weather. **Clarification Statement:** Emphasis is on local forms of severe weather.	**ESS3.B. Natural Hazards** Some kinds of severe weather are more likely than others in a given region. Weather scientists forecast severe weather so that the communities can prepare for and respond to these events. **ETS1.A. Defining and Delimiting Engineering Problems** Asking questions, making observations, and gathering information are helpful in thinking about problems. (K–2-ETS1-1)

Performance Expectations and Disciplinary Core Ideas for Kindergarten (*continued*)

Performance Expectations (PEs)	Disciplinary Core Ideas (DCIs)
K-ESS3-3. Communicate solutions that will reduce the impact of humans on the land, water, air, and/or other living things in the local environment. **Clarification Statement:** Examples of human impact on the land could include cutting trees to produce paper and using resources to produce bottles. Examples of solutions could include reusing paper and recycling cans and bottles.	**ESS3.C. Human Impacts on Earth Systems** Things that people do to live comfortably can affect the world around them. But they can make choices that reduce their impacts on the land, water, air, and other living things. **ETS1.B. Developing Possible Solutions** Designs can be conveyed through sketches, drawings, or physical models. These representations are useful in communicating ideas for a problem's solutions to other people. (2-LS2-2), (secondary to K-ESS3-3), (K–2-ETS1-2)
K-PS2-1. Plan and conduct an investigation to compare the effects of different strengths or different directions of pushes and pulls on the motion of an object. **Clarification Statement:** Examples of pushes or pulls could include a string attached to an object being pulled, a person pushing an object, a person stopping a rolling ball, and two objects colliding and pushing on each other. **Assessment Boundary:** Assessment is limited to different relative strengths or different directions, but not both at the same time. Assessment does not include non-contact pushes or pulls such as those produced by magnets.	**PS2.A. Forces and Motion** Pushes and pulls can have different strengths and directions. (K-PS2-2) Pushing or pulling on an object can change the speed or direction of its motion and can start or stop it. (K-PS2-2) **PS2.B. Types of Interactions** When objects touch or collide, they push on one another and can change motion. **PS3.C. Relationship Between Energy and Forces** A bigger push or pull makes things speed up or slow down more quickly.
K-PS2-2. Analyze data to determine if a design solution works as intended to change the speed or direction of an object with a push or a pull. **Clarification Statement:** Examples of problems requiring a solution could include having a marble or other object move a certain distance, follow a particular path, and knock down other objects. Examples of solutions could include tools such as a ramp to increase the speed of the object and a structure that would cause an object such as a marble or ball to turn. **Assessment Boundary:** Assessment does not include friction as a mechanism for change in speed.	**PS2.A. Forces and Motion** Pushes and pulls can have different strengths and directions. (K-PS2-1) **PS2.A. Forces and Motion** Pushing or pulling on an object can change the speed or direction of its motion and can start or stop it. (K-PS2-1) **ETS1.A. Defining and Delimiting Engineering Problems** A situation that people want to change or create can be approached as a problem to be solved through engineering. Such problems may have many acceptable solutions. (K–2-ETS1-1)
K-PS3-1. Make observations to determine the effect of sunlight on Earth's surface. **Clarification Statement:** Examples of Earth's surface could include sand, soil, rocks, and water. **Assessment Boundary:** Assessment of temperature is limited to relative measures such as warmer/cooler.	**PS3.B. Conservation of Energy and Energy Transfer** Sunlight warms Earth's surface. (K-PS3-1)
K-PS3-2. Use tools and materials provided to design and build a structure that will reduce the warming effect of sunlight on an area. **Clarification Statement:** Examples of structures could include umbrellas, canopies, and tents that minimize the warming effect of the Sun.	**PS3.B. Conservation of Energy and Energy Transfer** Sunlight warms Earth's surface. (K-PS3-2)

Performance Expectations and Disciplinary Core Ideas for Grade 1

Performance Expectations (PEs)	Disciplinary Core Ideas (DCIs)
1-LS1-1. Use materials to design a solution to a human problem by mimicking how plants and/or animals use their external parts to help them survive, grow, and meet their needs. **Clarification Statement:** Examples of human problems that can be solved by mimicking plant or animal solutions could include designing clothing or equipment to protect bicyclists by mimicking turtle shells, acorn shells, and animal scales; stabilizing structures by mimicking animal tails and roots on plants; keeping out intruders by mimicking thorns on branches and animal quills; and detecting intruders by mimicking eyes and ears.	**LS1.A. Structure and Function** All organisms have external parts. Different animals use their body parts in different ways to see, hear, grasp objects, protect themselves, move from place to place, and seek, find, and take in food, water, and air. Plants also have different parts (roots, stems, leaves, flowers, fruits) that help them survive and grow. **LS1.D. Information Processing** Animals have body parts that capture and convey different kinds of information needed for growth and survival. Animals respond to these inputs with behaviors that help them survive. Plants also respond to some external inputs.
1-LS1-2. Read texts and use media to determine patterns in behavior of parents and offspring that help offspring survive. **Clarification Statement:** Examples of patterns of behaviors could include the signals that offspring make (such as crying, cheeping, and other vocalizations) and the responses of the parents (such as feeding, comforting, and protecting the offspring).	**LS1.B. Growth and Development of Organisms** Adult plants and animals can have young. In many kinds of animals, parents and the offspring themselves engage in behaviors that help the offspring to survive.
1-LS3-1. Make observations to construct an evidence-based account that young plants and animals are like, but not exactly like, their parents. **Clarification Statement:** Examples of patterns could include features plants or animals share. Examples of observations could include that leaves from the same kind of plant are the same shape but can differ in size and that a particular breed of dog looks like its parents but is not exactly the same. **Assessment Boundary:** Assessment does not include inheritance or animals that undergo metamorphosis or hybrids.	**LS3.A. Inheritance of Traits** Young animals are very much, but not exactly like, their parents. Plants also are very much, but not exactly, like their parents. **LS3.B. Variation of Traits** Individuals of the same kind of plant or animal are recognizable as similar but can also vary in many ways.
1-ESS1-1. Use observations of the Sun, Moon, and stars to describe patterns that can be predicted. **Clarification Statement:** Examples of patterns could include that the Sun and Moon appear to rise in one part of the sky, move across the sky, and set; and stars other than our Sun are visible at night but not during the day. **Assessment Boundary:** Assessment of star patterns is limited to stars being seen at night and not during the day.	**ESS1.A. The Universe and Its Stars** Patterns of the motion of the Sun, Moon, and stars in the sky can be observed, described, and predicted.

Performance Expectations and Disciplinary Core Ideas for Grade 1 (*continued*)

Performance Expectations (PEs)	Disciplinary Core Ideas (DCIs)
1-ESS1-2. Make observations at different times of year to relate the amount of daylight to the time of year. **Clarification Statement:** Emphasis is on relative comparisons of the amount of daylight in the winter to the amount in the spring or fall. **Assessment Boundary:** Assessment is limited to relative amounts of daylight, not quantifying the hours or time of daylight.	**ESS1.B. Earth and the Solar System** Seasonal patterns of sunrise and sunset can be observed, described, and predicted.
1-PS4-1. Plan and conduct investigations to provide evidence that vibrating materials can make sound and that sound can make materials vibrate. **Clarification Statement:** Examples of vibrating materials that make sound could include tuning forks and plucking a stretched string. Examples of how sound can make matter vibrate could include holding a piece of paper near a speaker making sound and holding an object near a vibrating tuning fork.	**PS4.A. Wave Properties** Sound can make matter vibrate, and vibrating matter can make sound.
1-PS4-2. Make observations to construct an evidence-based account that objects in darkness can be seen only when illuminated. **Clarification Statement:** Examples of observations could include those made in a completely dark room, a pinhole box, and a video of a cave explorer with a flashlight. Illumination could be from an external light source or by an object giving off its own light.	**PS4.B. Electromagnetic Radiation** Objects can be seen if light is available to illuminate them or if they give off their own light.
1-PS4-3. Plan and conduct investigations to determine the effect of placing objects made with different materials in the path of a beam of light. **Clarification Statement:** Examples of materials could include those that are transparent (such as clear plastic), translucent (such as wax paper), opaque (such as cardboard), and reflective (such as a mirror). **Assessment Boundary:** Assessment does not include the speed of light.	**PS4.B. Electromagnetic Radiation** Some materials allow light to pass through them, others allow only some light through, and others block all the light and create a dark shadow on any surface beyond them, where the light cannot reach. Mirrors can be used to redirect a light beam. (Boundary: The idea that light travels from place to place is developed through experiences with light sources, mirrors, and shadows, but no attempt is made to discuss the speed of light.)
1-PS4-4. Use tools and materials to design and build a device that uses light or sound to solve the problem of communicating over a distance. **Clarification Statement:** Examples of devices could include a light source to send signals, paper cup and string "telephones," and a pattern of drum beats. **Assessment Boundary:** Assessment does not include technological details for how communication devices work.	**PS4.C. Information Technologies and Instrumentation** People use a variety of devices to communicate (send and receive information) over long distances.

Performance Expectations and Disciplinary Core Ideas for Grade 2

Performance Expectations (PEs)	Disciplinary Core Ideas (DCIs)
2-LS2-1. Plan and conduct an investigation to determine if plants need sunlight and water to grow. **Assessment Boundary:** Assessment is limited to testing one variable at a time.	**LS2.A. Interdependent Relationships in Ecosystems** Plants depend on water and light to grow.
2-LS2-2. Develop a simple model that mimics the function of an animal in dispersing seeds or pollinating plants.	**LS2.A. Interdependent Relationships in Ecosystems** Plants depend on animals for pollination or to move their seeds around. **ETS1.B. Developing Possible Solutions** Designs can be conveyed through sketches, drawings, or physical models. These representations are useful in communicating ideas for a problem's solutions to other people. (2-LS2-2), (K-ESS3-3), (K–2-ETS1-2)
2-LS4-1. Make observations of plants and animals to compare the diversity of life in different habitats. **Clarification Statement:** Emphasis is on the diversity of living things in each of a variety of different habitats. **Assessment Boundary:** Assessment does not include specific animal and plant names in specific habitats.	**LS4.D. Biodiversity and Humans** There are many different kinds of living things in any area, and they exist in different places on land and in water.
2-ESS1-1. Use information from several sources to provide evidence that Earth events can occur quickly or slowly. **Clarification Statement:** Examples of events and time scales could include volcanic explosions and earthquakes, which happen quickly and erosion of rocks, which occurs slowly. **Assessment Boundary:** Assessment does not include quantitative measurements of time scales.	**ESS1.C. The History of Planet Earth** Some events happen very quickly; others occur very slowly, over a time period much longer than one can observe.
2-ESS2-1. Compare multiple solutions designed to slow or prevent wind or water from changing the shape of the land. **Clarification Statement:** Examples of solutions could include different designs of dikes and windbreaks to hold back wind and water, and different designs for using shrubs, grass, and trees to hold back the land.	**ESS2.A. Earth Materials and Systems** Wind and water can change the shape of the land. **ETS1.C. Optimizing the Design Solution** Because there is always more than one possible solution to a problem, it is useful to compare and test designs. (K–2-ETS1-3)
2-ESS2-2. Develop a model to represent the shapes and kinds of land and bodies of water in an area. **Assessment Boundary:** Assessment does not include quantitative scaling in models.	**ESS2.B. Plate Tectonics and Large-Scale System Interactions** Maps show where things are located. One can map the shapes and kinds of land and water in any area.
2-ESS2-3. Obtain information to identify where water is found on Earth and that it can be solid or liquid.	**ESS2.C. The Roles of Water in Earth's Surface Processes** Water is found in the ocean, rivers, lakes, and ponds. Water exists as solid ice and in liquid form.
2-PS1-1. Plan and conduct an investigation to describe and classify different kinds of materials by their observable properties. **Clarification Statement:** Observations could include color, texture, hardness, and flexibility. Patterns could include the similar properties that different materials share.	**PS1.A. Structure and Properties of Matter** Different kinds of matter exist and many of them can be either solid or liquid, depending on temperature. Matter can be described and classified by its observable properties.

Performance Expectations and Disciplinary Core Ideas for Grade 2 (*continued*)

Performance Expectations (PEs)	Disciplinary Core Ideas (DCIs)
2-PS1-2. Analyze data obtained from testing different materials to determine which materials have the properties that are best suited for an intended purpose. **Clarification Statement:** Examples of properties could include strength, flexibility, hardness, texture, and absorbency. **Assessment Boundary:** Assessment of quantitative measurements is limited to length.	**PS1.A. Structure and Properties of Matter** Different properties are suited to different purposes. (2-PS1-3)
2-PS1-3. Make observations to construct an evidence-based account of how an object made of a small set of pieces can be disassembled and made into a new object. **Clarification Statement:** Examples of pieces could include blocks, building bricks, or other assorted small objects.	**PS1.A. Structure and Properties of Matter** Different properties are suited to different purposes. (2-PS1-2) A great variety of objects can be built up from a small set of pieces.
2-PS1-4. Construct an argument with evidence that some changes caused by heating or cooling can be reversed and some cannot. **Clarification Statement:** Examples of reversible changes could include materials such as water and butter at different temperatures. Examples of irreversible changes could include cooking an egg, freezing a plant leaf, and heating paper.	**PS1.B. Chemical Reactions** Heating or cooling a substance may cause changes that can be observed. Sometimes these changes are reversible, and sometimes they are not.

Performance Expectations and Disciplinary Core Ideas for Engineering Design in Grades K–2

Performance Expectations (PEs)	Disciplinary Core Ideas (DCIs)
K–2-ETS1-1. Ask questions, make observations, and gather information about a situation people want to change to define a simple problem that can be solved through the development of a new or improved object or tool.	**ETS1.A. Defining and Delimiting Engineering Problems** A situation that people want to change or create can be approached as a problem to be solved through engineering. Such problems may have many acceptable solutions. (K-PS2-2) **ETS1.A. Defining and Delimiting Engineering Problems** Asking questions, making observations, and gathering information are helpful in thinking about problems. (K-ESS3-2) **ETS1.A. Defining and Delimiting Engineering Problems** Before beginning to design a solution, it is important to clearly understand the problem.
K–2-ETS1-2. Develop a simple sketch, drawing, or physical model to illustrate how the shape of an object helps it function as needed to solve a given problem.	**ETS1.B. Developing Possible Solutions** Designs can be conveyed through sketches, drawings, or physical models. These representations are useful in communicating ideas for a problem's solutions to other people. (2-LS2-2), (K-ESS3-3), (K–2-ETS1-2)
K–2-ETS1-3. Analyze data from tests of two objects designed to solve the same problem to compare the strengths and weaknesses of how each performs.	**ETS1.C. Optimizing the Design Solution** Because there is always more than one possible solution to a problem, it is useful to compare and test designs. (2-ESS2-1)

CHAPTER 4

Focus on Grades 3–5

Science and Engineering Practices

Asking Questions and Defining Problems for Grades 3–5
Asking questions and defining problems in 3–5 builds on K–2 experiences and progresses to specifying qualitative relationships. • Ask questions about what would happen if a variable is changed. • Identify scientific (testable) and non-scientific (non-testable) questions. • Ask questions that can be investigated and predict reasonable outcomes based on patterns such as cause-and-effect relationships. • Use prior knowledge to describe problems that can be solved. • Define a simple design problem that can be solved through the development of an object, tool, process, or system and includes several criteria for success and constraints on materials, time, or cost.
Developing and Using Models for Grades 3–5
Modeling in 3–5 builds on K–2 experiences and progresses to building and revising simple models and using models to represent events and design solutions. • Identify limitations of models. • Collaboratively develop and/or revise a model based on evidence that shows the relationships among variables for frequent and regular occurring events. • Develop a model using an analogy, example, or abstract representation to describe a scientific principle or design solution. • Develop and/or use models to describe and/or predict phenomena. • Develop a diagram or simple physical prototype to convey a proposed object, tool, or process. • Use a model to test cause-and-effect relationships or interactions concerning the functioning of a natural or designed system.
Planning and Carrying Out Investigations for Grades 3–5
Planning and carrying out investigations to answer questions or test solutions to problems in 3–5 builds on K–2 experiences and progresses to include investigations that control variables and provide evidence to support explanations or design solutions. • Plan and conduct an investigation collaboratively to produce data to serve as the basis for evidence, using fair tests in which variables are controlled and the number of trials considered. • Evaluate appropriate methods and/or tools for collecting data. • Make observations and/or measurements to produce data to serve as the basis for evidence for an explanation of a phenomenon or test a design solution. • Make predictions about what would happen if a variable changes. • Test two different models of the same proposed object, tool, or process to determine which better meets criteria for success.
Analyzing and Interpreting Data for Grades 3–5
Analyzing data in 3–5 builds on K–2 experiences and progresses to introducing quantitative approaches to collecting data and conducting multiple trials of qualitative observations. When possible and feasible, digital tools should be used. • Represent data in tables and/or various graphical displays (bar graphs, pictographs, and/or pie charts) to reveal patterns that indicate relationships. • Analyze and interpret data to make sense of phenomena, using logical reasoning, mathematics, and/or computation. • Compare and contrast data collected by different groups in order to discuss similarities and differences in their findings. • Analyze data to refine a problem statement or the design of a proposed object, tool, or process. • Use data to evaluate and refine design solutions.

Science and Engineering Practices (*continued*)

Using Mathematics and Computational Thinking for Grades 3–5

Mathematical and computational thinking in 3–5 builds on K–2 experiences and progresses to extending quantitative measurements to a variety of physical properties and using computation and mathematics to analyze data and compare alternative design solutions.

- Organize simple data sets to reveal patterns that suggest relationships.
- Describe, measure, estimate, and/or graph quantities such as area, volume, weight, and time to address scientific and engineering questions and problems.
- Create and/or use graphs and/or charts generated from simple algorithms to compare alternative solutions to an engineering problem.

Constructing Explanations and Designing Solutions for Grades 3–5

Constructing explanations and designing solutions in 3–5 builds on K–2 experiences and progresses to the use of evidence in constructing explanations that specify variables that describe and predict phenomena and in designing multiple solutions to design problems.

- Construct an explanation of observed relationships (e.g., the distribution of plants in the backyard).
- Use evidence (e.g., measurements, observations, patterns) to construct or support an explanation or design a solution to a problem.
- Identify the evidence that supports particular points in an explanation.
- Apply scientific ideas to solve design problems.
- Generate and compare multiple solutions to a problem based on how well they meet the criteria and constraints of the design solution.

Engaging in Argument From Evidence for Grades 3–5

Engaging in argument from evidence in 3–5 builds on K–2 experiences and progresses to critiquing the scientific explanations or solutions proposed by peers by citing relevant evidence about the natural and designed world(s).

- Compare and refine arguments based on an evaluation of the evidence presented.
- Distinguish among facts, reasoned judgment based on research findings, and speculation in an explanation.
- Respectfully provide and receive critiques from peers about a proposed procedure, explanation, or model by citing relevant evidence and posing specific questions.
- Construct and/or support an argument with evidence, data, and/or a model.
- Use data to evaluate claims about cause and effect.
- Make a claim about the merit of a solution to a problem by citing relevant evidence about how it meets the criteria and constraints of the problem.

Obtaining, Evaluating, and Communicating Information for Grades 3–5

Obtaining, evaluating, and communicating information in 3–5 builds on K–2 experiences and progresses to evaluating the merit and accuracy of ideas and methods.

- Read and comprehend grade-appropriate complex texts and/or other reliable media to summarize and obtain scientific and technical ideas and describe how they are supported by evidence.
- Compare and/or combine across complex texts and/or other reliable media to support the engagement in other scientific and/or engineering practices.
- Combine information in written text with that contained in corresponding tables, diagrams, and/or charts to support the engagement in other scientific and/or engineering practices.
- Obtain and combine information from books and/or other reliable media to explain phenomena or solutions to a design problem.
- Communicate scientific and/or technical information orally and/or in written formats, including various forms of media as well as tables, diagrams, and charts.

Chapter 4

Crosscutting Concepts and Connections to Engineering, Technology, and Applications of Science

Crosscutting Concepts for Grades 3–5	
Patterns	• Similarities and differences in patterns can be used to sort, classify, communicate, and analyze simple rates of change for natural phenomena and designed products. • Patterns of change can be used to make predictions. • Patterns can be used as evidence to support an explanation.
Cause and Effect: Mechanism and Prediction	• Cause-and-effect relationships are routinely identified, tested, and used to explain change. • Events that occur together with regularity might or might not be a cause-and-effect relationship
Scale, Proportion, and Quantity	• Observable phenomena exist from very short to very long time periods. • Standard units are used to measure and describe physical quantities such as weight, time, temperature, and volume.
Systems and System Models	• A system is a group of related parts that make up a whole and can carry out functions its individual parts cannot. • A system can be described in terms of its components and their interactions.
Energy and Matter: Flows, Cycles, and Conservation	• Energy can be transferred in various ways and between objects. • Matter is made of particles. • Matter flows and cycles can be tracked in terms of the weight of the substances before and after a process occurs. The total weight of the substances does not change. This is what is meant by conservation of matter. Matter is transported into, out of, and within systems..
Structure and Function	• Different materials have different substructures, which can sometimes be observed. • Substructures have shapes and parts that serve functions
Stability and Change	• Change is measured in terms of differences over time and may occur at different rates. • Some systems appear stable, but over long periods of time will eventually change.
Connections to Engineering, Technology, and Applications of Science for Grades 3–5	
Interdependence of Science, Engineering, and Technology	• Science and technology support each other. • Tools and instruments are used to answer scientific questions, while scientific discoveries lead to the development of new technologies.
Influence of Science, Engineering, and Technology on Society and the Natural World	• People's needs and wants change over time, as do their demands for new and improved technologies. • Engineers improve existing technologies or develop new ones to increase their benefits, to decrease known risks, and to meet societal demands. • When new technologies become available, they can bring about changes in the way people live and interact with one another.

Connections to the Nature of Science

Understandings Most Closely Associated With Practices for Grades 3–5	
Scientific Investigations Use a Variety of Methods	• Science methods are determined by questions. • Science investigations use a variety of methods, tools, and techniques.
Scientific Knowledge Is Based on Empirical Evidence	• Science findings are based on recognizing patterns. • Science uses tools and technologies to make accurate measurements and observations.
Scientific Knowledge Is Open to Revision in Light of New Evidence	• Science explanations can change based on new evidence.
Science Models, Laws, Mechanisms, and Theories Explain Natural Phenomena	• Science theories are based on a body of evidence and many tests. • Science explanations describe the mechanisms for natural events.
Understandings Most Closely Associated With Crosscutting Concepts for Grades 3–5	
Science Is a Way of Knowing	• Science is both a body of knowledge and processes that add new knowledge. • Science is a way of knowing that is used by many people.
Scientific Knowledge Assumes an Order and Consistency in Natural Systems	• Science assumes consistent patterns in natural systems. • Basic laws of nature are the same everywhere in the universe.
Science Is a Human Endeavor	• Men and women from all cultures and backgrounds choose careers as scientists and engineers. • Most scientists and engineers work in teams. • Science affects everyday life. • Creativity and imagination are important to science.
Science Addresses Questions About the Natural and Material World	• Science findings are limited to questions that can be answered with empirical evidence.

Performance Expectations and Disciplinary Core Ideas for Grade 3

Performance Expectations (PEs)	Disciplinary Core Ideas (DCIs)
3-LS1-1. Develop models to describe that organisms have unique and diverse life cycles but all have in common birth, growth, reproduction, and death. **Clarification Statement:** Changes organisms go through during their life form a pattern. **Assessment Boundary:** Assessment of plant life cycles is limited to those of flowering plants. Assessment does not include details of human reproduction.	**LS1.B. Growth and Development of Organisms** Reproduction is essential to the continued existence of every kind of organism. Plants and animals have unique and diverse life cycles.
3-LS2-1. Construct an argument that some animals form groups that help members survive.	**LS2.D. Social Interactions and Group Behavior** Being part of a group helps animals obtain food, defend themselves, and cope with changes. Groups may serve different functions and vary dramatically in size.
3-LS3-1. Analyze and interpret data to provide evidence that plants and animals have traits inherited from parents and that variation of these traits exists in a group of similar organisms. **Clarification Statement:** Patterns are the similarities and differences in traits shared between offspring and their parents, or among siblings. Emphasis is on organisms other than humans. **Assessment Boundary:** Assessment does not include genetic mechanisms of inheritance and prediction of traits. Assessment is limited to non-human examples.	**LS3.A. Inheritance of Traits** Many characteristics of organisms are inherited from their parents. **LS3.B. Variation of Traits** Different organisms vary in how they look and function because they have different inherited information.
3-LS3-2. Use evidence to support the explanation that traits can be influenced by the environment. **Clarification Statement:** Examples of the environment affecting a trait could include that normally tall plants grown with insufficient water are stunted and that a pet dog that is given too much food and little exercise may become overweight.	**LS3.A. Inheritance of Traits** Other characteristics result from individuals' interactions with the environment, which can range from diet to learning. Many characteristics involve both inheritance and environment. **LS3.B. Variation of Traits** The environment also affects the traits that an organism develops.
3-LS4-1. Analyze and interpret data from fossils to provide evidence of the organisms and the environments in which they lived long ago. **Clarification Statement:** Examples of data could include type, size, and distributions of fossil organisms. Examples of fossils and environments could include marine fossils found on dry land, tropical plant fossils found in Arctic areas, and fossils of extinct organisms. **Assessment Boundary:** Assessment does not include identification of specific fossils or present plants and animals. Assessment is limited to major fossil types and relative ages.	**LS4.A. Evidence of Common Ancestry and Diversity** Some kinds of plants and animals that once lived on Earth are no longer found anywhere. Fossils provide evidence about the types of organisms that lived long ago and also about the nature of their environments.
3-LS4-2. Use evidence to construct an explanation for how the variations in characteristics among individuals of the same species may provide advantages in surviving, finding mates, and reproducing. **Clarification Statement:** Examples of cause-and-effect relationships could be that plants that have larger thorns than other plants may be less likely to be eaten by predators and that animals that have better camouflage coloration than other animals may be more likely to survive and therefore more likely to leave offspring.	**LS4.B. Natural Selection** Sometimes the differences in characteristics between individuals of the same species provide advantages in surviving, finding mates, and reproducing.

Performance Expectations and Disciplinary Core Ideas for Grade 3 (*continued*)

Performance Expectations (PEs)	Disciplinary Core Ideas (DCIs)
3-LS4-3. Construct an argument with evidence that in a particular habitat some organisms can survive well, some survive less well, and some cannot survive at all. **Clarification Statement:** Examples of evidence could include needs and characteristics of the organisms and habitats involved. The organisms and their habitat make up a system in which the parts depend on each other.	**LS4.C. Adaptation** For any particular environment, some kinds of organisms survive well, some survive less well, and some cannot survive at all.
3-LS4-4. Make a claim about the merit of a solution to a problem caused when the environment changes and the types of plants and animals that live there may change. **Clarification Statement:** Examples of environmental changes could include changes in land characteristics, water distribution, temperature, food, and other organisms. **Assessment Boundary:** Assessment is limited to a single environmental change. Assessment does not include the greenhouse effect or climate change.	**LS2.C. Ecosystem Dynamics, Functioning, and Resilience** When the environment changes in ways that affect a place's physical characteristics, temperature, or availability of resources, some organisms survive and reproduce, others move to new locations, yet others move into the transformed environment, and some die. **LS4.D. Biodiversity and Humans** Populations live in a variety of habitats, and change in those habitats affects the organisms living there.
3-ESS2-1. Represent data in tables and graphical displays to describe typical weather conditions expected during a particular season. **Clarification Statement:** Examples of data could include average temperature, precipitation, and wind direction. **Assessment Boundary:** Assessment of graphical displays is limited to pictographs and bar graphs. Assessment does not include climate change.	**ESS2.D. Weather and Climate** Scientists record patterns of the weather across different times and areas so that they can make predictions about what kind of weather might happen next.
3-ESS2-2. Obtain and combine information to describe climates in different regions of the world.	**ESS2.D. Weather and Climate** Climate describes a range of an area's typical weather conditions and the extent to which those conditions vary over years.
3-ESS3-1. Make a claim about the merit of a design solution that reduces the impacts of a weather-related hazard. **Clarification Statement:** Examples of design solutions to weather-related hazards could include barriers to prevent flooding, wind-resistant roofs, and lightning rods.	**ESS3.B. Natural Hazards** A variety of natural hazards result from natural processes. Humans cannot eliminate natural hazards but can take steps to reduce their impacts. (4-ESS3-2)
3-PS2-1. Plan and conduct an investigation to provide evidence of the effects of balanced and unbalanced forces on the motion of an object. **Clarification Statement:** Examples could include that an unbalanced force on one side of a ball can make it start moving and that balanced forces pushing on a box from both sides will not produce any motion at all. **Assessment Boundary:** Assessment is limited to one variable at a time: number, size, or direction of forces. Assessment does not include quantitative force size, only qualitative and relative. Assessment is limited to gravity being addressed as a force that pulls objects down.	**PS2.A. Forces and Motion** Each force acts on one particular object and has both strength and a direction. An object at rest typically has multiple forces acting on it, but they add to give zero net force on the object. Forces that do not sum to zero can cause changes in the object's speed or direction of motion. (Boundary: Qualitative and conceptual, but not quantitative addition of forces are used at this level.) **PS2.B. Types of Interactions** Objects in contact exert forces on each other.

Performance Expectations and Disciplinary Core Ideas for Grade 3 (*continued*)

Performance Expectations (PEs)	Disciplinary Core Ideas (DCIs)
3-PS2-2. Make observations and/or measurements of an object's motion to provide evidence that a pattern can be used to predict future motion. **Clarification Statement:** Examples of motion with a predictable pattern could include a child swinging in a swing, a ball rolling back and forth in a bowl, and two children on a see-saw. **Assessment Boundary:** Assessment does not include technical terms such as period and frequency.	**PS2.A. Forces and Motion** The patterns of an object's motion in various situations can be observed and measured; when that past motion exhibits a regular pattern, future motion can be predicted from it. (Boundary: Technical terms, such as magnitude, velocity, momentum, and vector quantity, are not introduced at this level, but the concept that some quantities need both size and direction to be described is developed.)
3-PS2-3. Ask questions to determine cause and effect relationships of electric or magnetic interactions between two objects not in contact with each other. **Clarification Statement:** Examples of an electric force could include the force on hair from an electrically charged balloon and the electrical forces between a charged rod and pieces of paper; examples of a magnetic force could include the force between two permanent magnets, the force between an electromagnet and steel paperclips, and the force exerted by one magnet versus the force exerted by two magnets. Examples of cause and effect relationships could include how the distance between objects affects strength of the force and how the orientation of magnets affects the direction of the magnetic force. **Assessment Boundary:** Assessment is limited to forces produced by objects that can be manipulated by students, and electrical interactions are limited to static electricity.	**PS2.B. Types of Interactions** Electric, and magnetic forces between a pair of objects do not require that the objects be in contact. The sizes of the forces in each situation depend on the properties of the objects and their distances apart and, for forces between two magnets, on their orientation relative to each other. (3-PS2-4)
3-PS2-4. Define a simple design problem that can be solved by applying scientific ideas about magnets. **Clarification Statement:** Examples of problems could include constructing a latch to keep a door shut and creating a device to keep two moving objects from touching each other.	**PS2.B. Types of Interactions** Electric, and magnetic forces between a pair of objects do not require that the objects be in contact. The sizes of the forces in each situation depend on the properties of the objects and their distances apart and, for forces between two magnets, on their orientation relative to each other. (3-PS2-3)

Performance Expectations and Disciplinary Core Ideas for Grade 4

Performance Expectations (PEs)	Disciplinary Core Ideas (DCIs)
4-LS1-1. Construct an argument that plants and animals have internal and external structures that function to support survival, growth, behavior, and reproduction. **Clarification Statement:** Examples of structures could include thorns, stems, roots, colored petals, heart, stomach, lung, brain, and skin. **Assessment Boundary:** Assessment is limited to macroscopic structures within plant and animal systems.	**LS1.A. Structure and Function** Plants and animals have both internal and external structures that serve various functions in growth, survival, behavior, and reproduction.
4-LS1-2. Use a model to describe that animals receive different types of information through their senses, process the information in their brain, and respond to the information in different ways. **Clarification Statement:** Emphasis is on systems of information transfer. **Assessment Boundary:** Assessment does not include the mechanisms by which the brain stores and recalls information or the mechanisms of how sensory receptors function.	**LS1.D. Information Processing** Different sense receptors are specialized for particular kinds of information, which may be then processed by the animal's brain. Animals are able to use their perceptions and memories to guide their actions.
4-ESS1-1. Identify evidence from patterns in rock formations and fossils in rock layers to support an explanation for changes in a landscape over time. **Clarification Statement:** Examples of evidence from patterns could include rock layers with marine shell fossils above rock layers with plant fossils and no shells, indicating a change from land to water over time; and a canyon with different rock layers in the walls and a river in the bottom, indicating that over time a river cut through the rock. **Assessment Boundary:** Assessment does not include specific knowledge of the mechanism of rock formation or memorization of specific rock formations and layers. Assessment is limited to relative time.	**ESS1.C. The History of Planet Earth** Local, regional, and global patterns of rock formations reveal changes over time due to Earth forces, such as earthquakes. The presence and location of certain fossil types indicate the order in which rock layers were formed.
4-ESS2-1. Make observations and/or measurements to provide evidence of the effects of weathering or the rate of erosion by water, ice, wind, or vegetation. **Clarification Statement:** Examples of variables to test could include angle of slope in the downhill movement of water, amount of vegetation, speed of wind, relative rate of deposition, cycles of freezing and thawing of water, cycles of heating and cooling, and volume of water flow. **Assessment Boundary:** Assessment is limited to a single form of weathering or erosion.	**ESS2.A. Earth Materials and Systems** Rainfall helps to shape the land and affects the types of living things found in a region. Water, ice, wind, living organisms, and gravity break rocks, soils, and sediments into smaller particles and move them around. **ESS2.E. Biogeology** Living things affect the physical characteristics of their regions.
4-ESS2-2. Analyze and interpret data from maps to describe patterns of Earth's features. **Clarification Statement:** Maps can include topographic maps of Earth's land and ocean floor, as well as maps of the locations of mountains, continental boundaries, volcanoes, and earthquakes.	**ESS2.B. Plate Tectonics and Large-Scale System Interactions** The locations of mountain ranges, deep ocean trenches, ocean floor structures, earthquakes, and volcanoes occur in patterns. Most earthquakes and volcanoes occur in bands that are often along the boundaries between continents and oceans. Major mountain chains form inside continents or near their edges. Maps can help locate the different land and water features of Earth.

Performance Expectations and Disciplinary Core Ideas for Grade 4 (*continued*)

Performance Expectations (PEs)	Disciplinary Core Ideas (DCIs)
4-ESS3-1. Obtain and combine information to describe that energy and fuels are derived from natural resources and their uses affect the environment. **Clarification Statement:** Examples of renewable energy resources could include wind energy, water behind dams, and sunlight; non-renewable energy resources are fossil fuels and fissile materials. Examples of environmental effects could include loss of habitat due to dams, loss of habitat due to surface mining, and air pollution from burning of fossil fuels.	**ESS3.A. Natural Resources** Energy and fuels that humans use are derived from natural sources, and their use affects the environment in multiple ways. Some resources are renewable over time, and others are not.
4-ESS3-2. Generate and compare multiple solutions to reduce the impacts of natural Earth processes on humans. **Clarification Statement:** Examples of solutions could include designing an earthquake-resistant building and improving monitoring of volcanic activity. **Assessment Boundary:** Assessment is limited to earthquakes, floods, tsunamis, and volcanic eruptions.	**ESS3.B. Natural Hazards** A variety of natural hazards result from natural processes. Humans cannot eliminate natural hazards but can take steps to reduce their impacts. (3-ESS3-1) **ETS1.B. Developing Possible Solutions** Testing a solution involves investigating how well it performs under a range of likely conditions.
4-PS3-1. Use evidence to construct an explanation relating the speed of an object to the energy of that object. **Assessment Boundary:** Assessment does not include quantitative measures of changes in the speed of an object or on any precise or quantitative definition of energy.	**PS3.A. Definitions of Energy** The faster a given object is moving, the more energy it possesses.
4-PS3-2. Make observations to provide evidence that energy can be transferred from place to place by sound, light, heat, and electric currents. **Assessment Boundary:** Assessment does not include quantitative measurements of energy.	**PS3.A. Definitions of Energy** Energy can be moved from place to place by moving objects or through sound, light, or electric currents. (4-PS3-3) **PS3.B. Conservation of Energy and Energy Transfer** Energy is present whenever there are moving objects, sound, light, or heat. When objects collide, energy can be transferred from one object to another, thereby changing their motion. In such collisions, some energy is typically also transferred to the surrounding air; as a result, the air gets heated and sound is produced. (4-PS3-3) Light also transfers energy from place to place. Energy can also be transferred from place to place by electric currents, which can then be used locally to produce motion, sound, heat, or light. The currents may have been produced to begin with by transforming the energy of motion into electrical energy. (4-PS3-4)

Performance Expectations and Disciplinary Core Ideas for Grade 4 (*continued*)

Performance Expectations (PEs)	Disciplinary Core Ideas (DCIs)
4-PS3-3. Ask questions and predict outcomes about the changes in energy that occur when objects collide. **Clarification Statement:** Emphasis is on the change in the energy due to the change in speed, not on the forces, as objects interact. **Assessment Boundary:** Assessment does not include quantitative measurements of energy.	**PS3.A. Definitions of Energy** Energy can be moved from place to place by moving objects or through sound, light, or electric currents. (4-PS3-2) **PS3.B. Conservation of Energy and Energy Transfer** Energy is present whenever there are moving objects, sound, light, or heat. When objects collide, energy can be transferred from one object to another, thereby changing their motion. In such collisions, some energy is typically also transferred to the surrounding air; as a result, the air gets heated and sound is produced. (4-PS3-2) **PS3.C. Relationship Between Energy and Forces** When objects collide, the contact forces transfer energy so as to change the objects' motions.
4-PS3-4. Apply scientific ideas to design, test, and refine a device that converts energy from one form to another. **Clarification Statement:** Examples of devices could include electric circuits that convert electrical energy into motion energy of a vehicle, light, or sound; and a passive solar heater that converts light into heat. Examples of constraints could include the materials, cost, or time to design the device. **Assessment Boundary:** Devices should be limited to those that convert motion energy to electric energy or use stored energy to cause motion or produce light or sound.	**PS3.B. Conservation of Energy and Energy Transfer** Energy can also be transferred from place to place by electric currents, which can then be used locally to produce motion, sound, heat, or light. The currents may have been produced to begin with by transforming the energy of motion into electrical energy. (4-PS3-2) **PS3.D. Energy in Chemical Processes and Everyday Life** The expression "produce energy" typically refers to the conversion of stored energy into a desired form for practical use. **ETS1.A. Defining and Delimiting Engineering Problems** Possible solutions to a problem are limited by available materials and resources (constraints). The success of a designed solution is determined by considering the desired features of a solution (criteria). Different proposals for solutions can be compared on the basis of how well each one meets the specified criteria for success or how well each takes the constraints into account. (3–5-ETS1-1)
4-PS4-1. Develop a model of waves to describe patterns in terms of amplitude and wavelength and that waves can cause objects to move. **Clarification Statement:** Examples of models could include diagrams, analogies, and physical models using wire to illustrate wavelength and amplitude of waves. **Assessment Boundary:** Assessment does not include interference effects, electromagnetic waves, non-periodic waves, or quantitative models of amplitude and wavelength.	**PS4.A. Wave Properties** Waves, which are regular patterns of motion, can be made in water by disturbing the surface. When waves move across the surface of deep water, the water goes up and down in place; there is no net motion in the direction of the wave except when the water meets a beach. **PS4.A. Wave Properties** Waves of the same type can differ in amplitude (height of the wave) and wavelength (spacing between wave peaks).

Performance Expectations and Disciplinary Core Ideas for Grade 4 (*continued*)

Performance Expectations (PEs)	Disciplinary Core Ideas (DCIs)
4-PS4-2. Develop a model to describe that light reflecting from objects and entering the eye allows objects to be seen. **Assessment Boundary:** Assessment does not include knowledge of specific colors reflected and seen, the cellular mechanisms of vision, or how the retina works.	**PS4.B. Electromagnetic Radiation** An object can be seen when light reflected from its surface enters the eyes.
4-PS4-3. Generate and compare multiple solutions that use patterns to transfer information. **Clarification Statement:** Examples of solutions could include drums sending coded information through sound waves, using a grid of 1's and 0's representing black and white to send information about a picture, and using Morse code to send text.	**PS4.C. Information Technologies and Instrumentation** Digitized information can be transmitted over long distances without significant degradation. High-tech devices, such as computers or cell phones, can receive and decode information—convert it from digitized form to voice—and vice versa. **ETS1.C. Optimizing the Design Solution** Different solutions need to be tested in order to determine which of them best solves the problem, given the criteria and the constraints. (3–5-ETS1-3)

Performance Expectations and Disciplinary Core Ideas for Grade 5

Performance Expectations (PEs)	Disciplinary Core Ideas (DCIs)
5-LS1-1. Support an argument that plants get the materials they need for growth chiefly from air and water. **Clarification Statement:** Emphasis is on the idea that plant matter comes mostly from air and water, not from the soil.	**LS1.C. Organization for Matter and Energy Flow in Organisms** Plants acquire their material for growth chiefly from air and water.
5-LS2-1. Develop a model to describe the movement of matter among plants, animals, decomposers, and the environment. **Clarification Statement:** Emphasis is on the idea that matter that is not food (air, water, decomposed materials in soil) is changed by plants into matter that is food. Examples of systems could include organisms, ecosystems, and the Earth. **Assessment Boundary:** Assessment does not include molecular explanations.	**LS2.A. Interdependent Relationships in Ecosystems** The food of almost any kind of animal can be traced back to plants. Organisms are related in food webs in which some animals eat plants for food and other animals eat the animals that eat plants. Some organisms, such as fungi and bacteria, break down dead organisms (both plants or plant parts and animals) and therefore operate as "decomposers." Decomposition eventually restores (recycles) some materials back to the soil. Organisms can survive only in environments in which their particular needs are met. A healthy ecosystem is one in which multiple species of different types are each able to meet their needs in a relatively stable web of life. Newly introduced species can damage the balance of an ecosystem. **LS2.B. Cycles of Matter and Energy Transfer in Ecosystems** Matter cycles between the air and soil and among plants, animals, and microbes as these organisms live and die. Organisms obtain gases and water from the environment and release waste matter (gas, liquid, or solid) back into the environment.
5-ESS1-1. Support an argument that differences in the apparent brightness of the Sun compared to other stars is due to their relative distances from the Earth. **Assessment Boundary:** Assessment is limited to relative distances, not sizes, of stars. Assessment does not include other factors that affect apparent brightness (such as stellar masses, age, stage).	**ESS1.A. The Universe and Its Stars** The Sun is a star that appears larger and brighter than other stars because it is closer. Stars range greatly in their distance from Earth.
5-ESS1-2. Represent data in graphical displays to reveal patterns of daily changes in length and direction of shadows, day and night, and the seasonal appearance of some stars in the night sky. **Clarification Statement:** Examples of patterns could include the position and motion of Earth with respect to the Sun and selected stars that are visible only in particular months. **Assessment Boundary:** Assessment does not include causes of seasons.	**ESS1.B. Earth and the Solar System** The orbits of Earth around the Sun and of the Moon around Earth, together with the rotation of Earth about an axis between its North and South poles, cause observable patterns. These include day and night; daily changes in the length and direction of shadows; and different positions of the Sun, Moon, and stars at different times of the day, month, and year.

Performance Expectations and Disciplinary Core Ideas for Grade 5 (*continued*)

Performance Expectations (PEs)	Disciplinary Core Ideas (DCIs)
5-ESS2-1. Develop a model using an example to describe ways the geosphere, biosphere, hydrosphere, and/or atmosphere interact. **Clarification Statement:** Examples could include the influence of the ocean on ecosystems, landform shape, and climate; the influence of the atmosphere on landforms and ecosystems through weather and climate; and the influence of mountain ranges on winds and clouds in the atmosphere. The geosphere, hydrosphere, atmosphere, and biosphere are each a system. **Assessment Boundary:** Assessment is limited to the interactions of two systems at a time.	**ESS2.A. Earth Materials and Systems** Earth's major systems are the geosphere (solid and molten rock, soil, and sediments), the hydrosphere (water and ice), the atmosphere (air), and the biosphere (living things, including humans). These systems interact in multiple ways to affect Earth's surface materials and processes. The ocean supports a variety of ecosystems and organisms, shapes landforms, and influences climate. Winds and clouds in the atmosphere interact with the landforms to determine patterns of weather.
5-ESS2-2. Describe and graph the amounts and percentages of water and fresh water in various reservoirs to provide evidence about the distribution of water on Earth. **Assessment Boundary:** Assessment is limited to oceans, lakes, rivers, glaciers, groundwater, and polar ice caps, and does not include the atmosphere.	**ESS2.C. The Roles of Water in Earth's Surface Processes** Nearly all of Earth's available water is in the ocean. Most fresh water is in glaciers or underground; only a tiny fraction is in streams, lakes, wetlands, and the atmosphere.
5-ESS3-1. Obtain and combine information about ways individual communities use science ideas to protect the Earth's resources and environment.	**ESS3.C. Human Impacts on Earth Systems** Human activities in agriculture, industry, and everyday life have had major effects on the land, vegetation, streams, ocean, air, and even outer space. But individuals and communities are doing things to help protect Earth's resources and environments.
5-PS1-1. Develop a model to describe that matter is made of particles too small to be seen. **Clarification Statement:** Examples of evidence could include adding air to expand a basketball, compressing air in a syringe, dissolving sugar in water, and evaporating salt water. **Assessment Boundary:** Assessment does not include the atomic-scale mechanism of evaporation and condensation or defining the unseen particles.	**PS1.A. Structure and Properties of Matter** Matter of any type can be subdivided into particles that are too small to see, but even then the matter still exists and can be detected by other means. A model showing that gases are made from matter particles that are too small to see and are moving freely around in space can explain many observations, including the inflation and shape of a balloon and the effects of air on larger particles or objects.
5-PS1-2. Measure and graph quantities to provide evidence that regardless of the type of change that occurs when heating, cooling, or mixing substances, the total weight of matter is conserved. **Clarification Statement:** Examples of reactions or changes could include phase changes, dissolving, and mixing that form new substances. **Assessment Boundary:** Assessment does not include distinguishing mass and weight.	**PS1.A. Structure and Properties of Matter** The amount (weight) of matter is conserved when it changes form, even in transitions in which it seems to vanish. **PS1.B. Chemical Reactions** No matter what reaction or change in properties occurs, the total weight of the substances does not change. (Boundary: Mass and weight are not distinguished at this grade level.)

Performance Expectations and Disciplinary Core Ideas for Grade 5 (*continued*)

Performance Expectations (PEs)	Disciplinary Core Ideas (DCIs)
5-PS1-3. Make observations and measurements to identify materials based on their properties. **Clarification Statement:** Examples of materials to be identified could include baking soda and other powders, metals, minerals, and liquids. Examples of properties could include color, hardness, reflectivity, electrical conductivity, thermal conductivity, response to magnetic forces, and solubility; density is not intended as an identifiable property. **Assessment Boundary:** Assessment does not include density or distinguishing mass and weight.	**PS1.A. Structure and Properties of Matter** Measurements of a variety of properties can be used to identify materials. (Boundary: At this grade level, mass and weight are not distinguished, and no attempt is made to define the unseen particles or explain the atomic-scale mechanism of evaporation and condensation.)
5-PS1-4. Conduct an investigation to determine whether the mixing of two or more substances results in new substances.	**PS1.B. Chemical Reactions** When two or more different substances are mixed, a new substance with different properties may be formed.
5-PS2-1. Support an argument that the gravitational force exerted by Earth on objects is directed down. **Clarification Statement:** "Down" is a local description of the direction that points toward the center of the spherical Earth. **Assessment Boundary:** Assessment does not include mathematical representation of gravitational force.	**PS2.B. Types of Interactions** The gravitational force of Earth acting on an object near Earth's surface pulls that object toward the planet's center.
5-PS3-1. Use models to describe that energy in animals' food (used for body repair, growth, motion, and to maintain body warmth) was once energy from the Sun. **Clarification Statement:** Examples of models could include diagrams and flow charts.	**LS1.C. Organization for Matter and Energy Flow in Organisms** Food provides animals with the materials they need for body repair and growth and the energy they need to maintain body warmth and for motion. **PS3.D. Energy in Chemical Processes and Everyday Life** The energy released from food was once energy from the Sun that was captured by plants in the chemical process that forms plant matter (from air and water).

Chapter 4

Performance Expectations and Disciplinary Core Ideas for Engineering Design in Grades 3–5

Performance Expectations (PEs)	Disciplinary Core Ideas (DCIs)
3-5-ETS1-1. Define a simple design problem reflecting a need or a want that includes specified criteria for success and constraints on materials, time, or cost.	**ETS1.A. Defining and Delimiting Engineering Problems** Possible solutions to a problem are limited by available materials and resources (constraints). The success of a designed solution is determined by considering the desired features of a solution (criteria). Different proposals for solutions can be compared on the basis of how well each one meets the specified criteria for success or how well each takes the constraints into account. (4-PS3-4)
3-5-ETS1-2. Generate and compare multiple possible solutions to a problem based on how well each is likely to meet the criteria and constraints of the problem.	**ETS1.B. Developing Possible Solutions** Research on a problem should be carried out before beginning to design a solution. Testing a solution involves investigating how well it performs under a range of likely conditions. **ETS1.B. Developing Possible Solutions** At whatever stage, communicating with peers about proposed solutions is an important part of the design process, and shared ideas can lead to improved designs.
3-5-ETS1-3. Plan and carry out fair tests in which variables are controlled and failure points are considered to identify aspects of a model or prototype that can be improved.	**ETS1.B. Developing Possible Solutions** Tests are often designed to identify failure points or difficulties, which suggest the elements of the design that need to be improved. **ETS1.C. Optimizing the Design Solution** Different solutions need to be tested in order to determine which of them best solves the problem, given the criteria and the constraints. (4-PS4-3)

CHAPTER 5

Focus on Middle School

Science and Engineering Practices

Asking Questions and Defining Problems for Grades 6–8

Asking questions and defining problems in 6–8 builds on K–5 experiences and progresses to specifying relationships between variables and clarifying arguments and models.

- Ask questions that arise from careful observation of phenomena, models, or unexpected results, to clarify and/or seek additional information.
- Ask questions to identify and/or clarify evidence and/or the premise(s) of an argument.
- Ask questions to determine relationships between independent and dependent variables and relationships in models.
- Ask questions to clarify and/or refine a model, an explanation, or an engineering problem.
- Ask questions that can be investigated within the scope of the classroom, outdoor environment, and museums and other public facilities with available resources and, when appropriate, frame a hypothesis based on observations and scientific principles.
- Define a design problem that can be solved through the development of an object, tool, process, or system and includes multiple criteria and constraints, including scientific knowledge that may limit possible solutions.

Developing and Using Models for Grades 6–8

Modeling in 6–8 builds on K–5 experiences and progresses to developing, using, and revising models to describe, test, and predict more abstract phenomena and design systems.

- Evaluate limitations of a model for a proposed object or tool.
- Develop or modify a model—based on evidence – to match what happens if a variable or component of a system is changed.
- Use and/or develop a model of simple systems with uncertain and less predictable factors.
- Develop and/or revise a model to show the relationships among variables, including those that are not observable but predict observable phenomena.
- Develop and/or use a model to predict and/or describe phenomena.
- Develop a model to describe unobservable mechanisms.
- Develop and/or use a model to generate data to test ideas about phenomena in natural or designed systems, including those representing inputs and outputs, and those at unobservable scales.

Planning and Carrying Out Investigations for Grades 6–8

Planning and carrying out investigations to answer questions or test colutions in 6–8 builds on K–5 experiences and progresses to include investigations that use multiple variables and provide evidence to support explanations or design solutions.

- Plan an investigation individually and collaboratively, and in the design identify independent and dependent variables and controls, what tools are needed to do the gathering, how measurements will be recorded, and how many data are needed to support a claim.
- Conduct an investigation and/or evaluate and/or revise the experimental design to produce data to serve as the basis for evidence that meet the goals of the investigation.
- Evaluate the accuracy of various methods for collecting data.
- Collect data to produce data to serve as the basis for evidence to answer scientific questions or test design solutions under a range of conditions.
- Collect data about the performance of a proposed object, tool, process, or system under a range of conditions.

Science and Engineering Practices (*continued*)

Analyzing and Interpreting Data for Grades 6–8

Analyzing data in 6–8 builds on K–5 experiences and progresses to extending quantitative analysis to investigations, distinguishing between correlation and causation, and basic statistical techniques of data and error analysis.

- Construct, analyze, and/or interpret graphical displays of data and/or large data sets to identify linear and nonlinear relationships.
- Use graphical displays (e.g., maps, charts, graphs, and/or tables) of large data sets to identify temporal and spatial relationships.
- Distinguish between causal and correlational relationships in data.
- Analyze and interpret data to provide evidence for phenomena.
- Apply concepts of statistics and probability (including mean, median, mode, and variability) to analyze and characterize data, using digital tools when feasible.
- Consider limitations of data analysis (e.g., measurement error), and/or seek to improve precision and accuracy of data with better technological tools and methods (e.g., multiple trials).
- Analyze and interpret data to determine similarities and differences in findings.
- Analyze data to define an optimal operational range for a proposed object, tool, process, or system that best meets criteria for success.

Using Mathematics and Computational Thinking for Grades 6–8

Mathematical and computational thinking in 6–8 builds on K–5 experiences and progresses to identifying patterns in large data sets and using mathematical concepts to support explanations and arguments.

- Decide when to use qualitative vs. quantitative data.
- Use digital tools (e.g., computers) to analyze very large data sets for patterns and trends.
- Use mathematical representations to describe and/or support scientific conclusions and design solutions.
- Create algorithms (a series of ordered steps) to solve a problem.
- Apply mathematical concepts and/or processes (such as ratio, rate, percent, basic operations, and simple algebra) to scientific and engineering questions and problems.
- Use digital tools and/or mathematical concepts and arguments to test and compare proposed solutions to an engineering design problem.

Constructing Explanations and Designing Solutions for Grades 6–8

Constructing explanations and designing solutions in 6–8 builds on K–5 experiences and progresses to include constructing explanations and designing solutions supported by multiple sources of evidence consistent with scientific ideas, principles, and theories.

- Construct an explanation that includes qualitative or quantitative relationships between variables that predict(s) and/or describe(s) phenomena.
- Construct an explanation using models or representations.
- Construct a scientific explanation based on valid and reliable evidence obtained from sources (including the students' own experiments) and the assumption that theories and laws that describe the natural world operate today as they did in the past and will continue to do so in the future.
- Apply scientific ideas, principles, and/or evidence to construct, revise, and/or use an explanation for real-world phenomena, examples, or events.
- Apply scientific reasoning to show why the data or evidence is adequate for the explanation or conclusion.
- Apply scientific ideas or principles to design, construct, and/or test a design of an object, tool, process, or system.
- Undertake a design project, engaging in the design cycle, to construct and/or implement a solution that meets specific design criteria and constraints.
- Optimize performance of a design by prioritizing criteria, making trade-offs, testing, revising, and retesting.

Science and Engineering Practices (*continued*)

Engaging in Argument From Evidence for Grades 6–8
Engaging in argument from evidence in 6–8 builds on K–5 experiences and progresses to constructing a convincing argument that supports or refutes claims for either explanations or solutions about the natural and designed world(s). • Compare and critique two arguments on the same topic and analyze whether they emphasize similar or different evidence and/or interpretations of facts. • Respectfully provide and receive critiques about one's explanations, procedures, models, and questions by citing relevant evidence and posing and responding to questions that elicit pertinent elaboration and detail. • Construct, use, and/or present an oral and written argument supported by empirical evidence and scientific reasoning to support or refute an explanation or a model for a phenomenon or a solution to a problem. • Make an oral or written argument that supports or refutes the advertised performance of a device, process, or system, based on empirical evidence concerning whether or not the technology meets relevant criteria and constraints. • Evaluate competing design solutions based on jointly developed and agreed-upon design criteria.
Obtaining, Evaluating, and Communicating Information for Grades 6–8
Obtaining, evaluating, and communicating information in 6–8 builds on K–5 experiences and progresses to evaluating the merit and validity of ideas and methods. • Critically read scientific texts adapted for classroom use to determine the central ideas and/or obtain scientific and/or technical information to describe patterns in and/or evidence about the natural and designed world(s). • Integrate qualitative and/or quantitative scientific and/or technical information in written text with that contained in media and visual displays to clarify claims and findings. • Gather, read, and synthesize information from multiple appropriate sources and assess the credibility, accuracy, and possible bias of each publication and methods used, and describe how they are supported or not supported by evidence. • Evaluate data, hypotheses, and/or conclusions in scientific and technical texts in light of competing information or accounts. • Communicate scientific and/or technical information (e.g., about a proposed object, tool, process, system) in writing and/or through oral presentations.

Crosscutting Concepts and Connections to Engineering, Technology, and Applications of Science

Crosscutting Concepts for Grades 6–8	
Patterns	• Macroscopic patterns are related to the nature of microscopic and atomic-level structure. • Patterns in rates of change and other numerical relationships can provide information about natural and human-designed systems. • Patterns can be used to identify cause-and-effect relationships. • Graphs, charts, and images can be used to identify patterns in data.
Cause and Effect: Mechanism and Prediction	• Relationships can be classified as causal or correlational, and correlation does not necessarily imply causation. • Cause-and-effect relationships may be used to predict phenomena in natural or designed systems. • Phenomena may have more than one cause, and some cause-and-effect relationships in systems can only be described using probability.
Scale, Proportion, and Quantity	• Time, space, and energy phenomena can be observed at various scales using models to study systems that are too large or too small. • The observed function of natural and designed systems may change with scale. • Proportional relationships (e.g., speed as the ratio of distance traveled to time taken) among different types of quantities provide information about the magnitude of properties and processes. • Scientific relationships can be represented through the use of algebraic expressions and equations. • Phenomena that can be observed at one scale may not be observable at another scale.
Systems and System Models	• Systems may interact with other systems; they may have subsystems and be a part of larger complex systems. • Models can be used to represent systems and their interactions—such as inputs, processes, and outputs—and energy, matter, and information flows within systems. • Models are limited in that they only represent certain aspects of the system under study.
Energy and Matter: Flows, Cycles, and Conservation	• Matter is conserved because atoms are conserved in physical and chemical processes. • Within a natural or designed system, the transfer of energy drives the motion and/or cycling of matter. • Energy may take different forms (e.g., energy in fields, thermal energy, energy of motion). • The transfer of energy can be tracked as energy flows through a designed or natural system.
Structure and Function	• Complex and microscopic structures and systems can be visualized, modeled, and used to describe how their function depends on the shapes, composition, and relationships among its parts; therefore, complex natural and designed structures/systems can be analyzed to determine how they function. • Structures can be designed to serve particular functions by taking into account properties of different materials, and how materials can be shaped and used.

Crosscutting Concepts and Connections to Engineering, Technology, and Applications of Science (*continued*)

Crosscutting Concepts for Grades 6–8	
Stability and Change	• Explanations of stability and change in natural or designed systems can be constructed by examining the changes over time and processes at different scales, including the atomic scale. • Small changes in one part of a system might cause large changes in another part. • Stability might be disturbed either by sudden events or gradual changes that accumulate over time. • Systems in dynamic equilibrium are stable due to a balance of feedback mechanisms.
Connections to Engineering, Technology, and Applications of Science for Grades 6–8	
Interdependence of Science, Engineering, and Technology	• Engineering advances have led to important discoveries in virtually every field of science, and scientific discoveries have led to the development of entire industries and engineered systems. • Science and technology drive each other forward.
Influence of Science, Engineering, and Technology on Society and the Natural World	• All human activity draws on natural resources and has both short- and long-term consequences, positive as well as negative, for the health of people and the natural environment. • The uses of technologies and any limitations on their use are driven by individual or societal needs, desires, and values; by the findings of scientific research; and by differences in such factors as climate, natural resources, and economic conditions. • Technology use varies over time and from region to region.

Connections to the Nature of Science

Understandings Most Closely Associated With Practices for Grades 6–8	
Scientific Investigations Use a Variety of Methods	• Science investigations use a variety of methods and tools to make measurements and observations. • Science investigations are guided by a set of values to ensure accuracy of measurements, observations, and objectivity of findings. • Science depends on evaluating proposed explanations. • Scientific values function as criteria in distinguishing between science and non-science.
Scientific Knowledge Is Based on Empirical Evidence	• Science knowledge is based upon logical and conceptual connections between evidence and explanations. • Science disciplines share common rules of obtaining and evaluating empirical evidence.
Scientific Knowledge Is Open to Revision in Light of New Evidence	• Scientific explanations are subject to revision and improvement in light of new evidence. • The certainty and durability of science findings vary. • Science findings are frequently revised and/or reinterpreted based on new evidence.
Science Models, Laws, Mechanisms, and Theories Explain Natural Phenomena	• Theories are explanations for observable phenomena. • Science theories are based on a body of evidence developed over time. • Laws are regularities or mathematical descriptions of natural phenomena. • A hypothesis is used by scientists as an idea that may contribute important new knowledge for the evaluation of a scientific theory. • The term "theory" as used in science is very different from the common use outside of science.
Understandings Most Closely Associated With Crosscutting Concepts for Grades 6–8	
Science Is a Way of Knowing	• Science is both a body of knowledge and the processes and practices used to add to that body of knowledge. • Science knowledge is cumulative and many people, from many generations and nations, have contributed to science knowledge. • Science is a way of knowing used by many people, not just scientists.
Scientific Knowledge Assumes an Order and Consistency in Natural Systems	• Science assumes that objects and events in natural systems occur in consistent patterns that are understandable through measurement and observation. • Science carefully considers and evaluates anomalies in data and evidence.
Science Is a Human Endeavor	• Men and women from different social, cultural, and ethnic backgrounds work as scientists and engineers. • Scientists and engineers rely on human qualities such as persistence, precision, reasoning, logic, imagination, and creativity. • Scientists and engineers are guided by habits of mind such as intellectual honesty, tolerance of ambiguity, skepticism, and openness to new ideas. • Advances in technology influence the progress of science, and science has influenced advances in technology.
Science Addresses Questions About the Natural and Material World	• Scientific knowledge is constrained by human capacity, technology, and materials. • Science limits its explanations to systems that lend themselves to observation and empirical evidence. • Science knowledge can describe consequences of actions but is not responsible for society's decisions.

Performance Expectations and Disciplinary Core Ideas for Physical Science

Performance Expectations (PEs)	Disciplinary Core Ideas (DCIs)
MS-PS1-1. Develop models to describe the atomic composition of simple molecules and extended structures. **Clarification Statement:** Emphasis is on developing models of molecules that vary in complexity. Examples of simple molecules could include ammonia and methanol. Examples of extended structures could include sodium chloride or diamonds. Examples of molecular-level models could include drawings, 3-D ball-and-stick structures, or computer representations showing different molecules with different types of atoms. **Assessment Boundary:** Assessment does not include valence electrons and bonding energy. Discussing the ionic nature of subunits of complex structures or a complete description of all individual atoms in a complex molecule or extended structure is not required.	**PS1.A. Structure and Properties of Matter** Substances are made from different types of atoms, which combine with one another in various ways. Atoms form molecules that range in size from two to thousands of atoms. Solids may be formed from molecules, or they may be extended structures with repeating subunits (e.g., crystals).
MS-PS1-2. Analyze and interpret data on the properties of substances before and after the substances interact to determine if a chemical reaction has occurred. **Clarification Statement:** Examples of reactions could include burning sugar or steel wool, fat reacting with sodium hydroxide, and mixing zinc with hydrogen chloride. **Assessment Boundary:** Assessment is limited to analysis of the following properties: density, melting point, boiling point, solubility, flammability, and odor.	**PS1.A. Structure and Properties of Matter** Each pure substance has characteristic physical and chemical properties (for any bulk quantity under given conditions) that can be used to identify it. (MS-PS1-3) **PS1.B. Chemical Reactions** Substances react chemically in characteristic ways. In a chemical process, the atoms that make up the original substances are regrouped into different molecules, and these new substances have different properties from those of the reactants. (MS-PS1-3), (MS-PS1-5)
MS-PS1-3. Gather and make sense of information to describe that synthetic materials come from natural resources and impact society. **Clarification Statement:** Emphasis is on natural resources that undergo a chemical process to form the synthetic material. Examples of new materials could include new medicine, foods, and alternative fuels. **Assessment Boundary:** Assessment is limited to qualitative information.	**PS1.A. Structure and Properties of Matter** Each pure substance has characteristic physical and chemical properties (for any bulk quantity under given conditions) that can be used to identify it. (MS-PS1-2) **PS1.B. Chemical Reactions** Substances react chemically in characteristic ways. In a chemical process, the atoms that make up the original substances are regrouped into different molecules, and these new substances have different properties from those of the reactants. (MS-PS1-2), (MS-PS1-5)

Performance Expectations and Disciplinary Core Ideas for Physical Science (*continued*)

Performance Expectations (PEs)	Disciplinary Core Ideas (DCIs)
MS-PS1-4. Develop a model that predicts and describes changes in particle motion, temperature, and state of a pure substance when thermal energy is added or removed. **Clarification Statement:** Emphasis is on qualitative molecular-level models of solids, liquids, and gases to show that adding or removing thermal energy increases or decreases kinetic energy of the particles until a change of state occurs. Examples of models could include drawing and diagrams. Examples of particles could include molecules or inert atoms. Examples of pure substances could include water, carbon dioxide, and helium.	**PS1.A. Structure and Properties of Matter** Gases and liquids are made of molecules or inert atoms that are moving about relative to each other. In a liquid, the molecules are constantly in contact with others; in a gas, they are widely spaced except when they happen to collide. In a solid, atoms are closely spaced and may vibrate in position but do not change relative locations. The changes of state that occur with variations in temperature or pressure can be described and predicted using these models of matter. **PS3.A. Definitions of Energy** The term "heat" as used in everyday language refers both to thermal energy (the motion of atoms or molecules within a substance) and the transfer of that thermal energy from one object to another. In science, heat is used only for this second meaning; it refers to the energy transferred due to the temperature difference between two objects. The temperature of a system is proportional to the average internal kinetic energy and potential energy per atom or molecule (whichever is the appropriate building block for the system's material). The details of that relationship depend on the type of atom or molecule and the interactions among the atoms in the material. Temperature is not a direct measure of a system's total thermal energy. The total thermal energy (sometimes called the total internal energy) of a system depends jointly on the temperature, the total number of atoms in the system, and the state of the material.
MS-PS1-5. Develop and use a model to describe how the total number of atoms does not change in a chemical reaction and thus mass is conserved. **Clarification Statement:** Emphasis is on law of conservation of matter and on physical models or drawings, including digital forms, that represent atoms. **Assessment Boundary:** Assessment does not include the use of atomic masses, balancing symbolic equations, or intermolecular forces.	**PS1.B. Chemical Reactions** Substances react chemically in characteristic ways. In a chemical process, the atoms that make up the original substances are regrouped into different molecules, and these new substances have different properties from those of the reactants. (MS-PS1-2), (MS-PS1-3) The total number of each type of atom is conserved, and thus the mass does not change.

Performance Expectations and Disciplinary Core Ideas for Physical Science (*continued*)

Performance Expectations (PEs)	Disciplinary Core Ideas (DCIs)
MS-PS1-6. Undertake a design project to construct, test, and modify a device that either releases or absorbs thermal energy by chemical processes. **Clarification Statement:** Emphasis is on the design, controlling the transfer of energy to the environment, and modification of a device using factors such as type and concentration of a substance. Examples of designs could involve chemical reactions such as dissolving ammonium chloride or calcium chloride. **Assessment Boundary:** Assessment is limited to the criteria of amount, time, and temperature of substance in testing the device.	**PS1.B. Chemical Reactions** Some chemical reactions release energy, others store energy. **ETS1.B. Developing Possible Solutions** A solution needs to be tested, and then modified on the basis of the test results in order to improve it. **ETS1.C. Optimizing the Design Solution** Although one design may not perform the best across all tests, identifying the characteristics of the design that performed the best in each test can provide useful information for the redesign process—that is, some of the characteristics may be incorporated into the new design. The iterative process of testing the most promising solutions and modifying what is proposed on the basis of the test results leads to greater refinement and ultimately to an optimal solution. (MS-ETS-4)
MS-PS2-1. Apply Newton's third law to design a solution to a problem involving the motion of two colliding objects. **Clarification Statement:** Examples of practical problems could include the impact of collisions between two cars, between a car and stationary objects, and between a meteor and a space vehicle. **Assessment Boundary:** Assessment is limited to vertical or horizontal interactions in one dimension.	**PS2.A. Forces and Motion** For any pair of interacting objects, the force exerted by the first object on the second object is equal in strength to the force that the second object exerts on the first, but in the opposite direction (Newton's third law).
MS-PS2-2. Plan an investigation to provide evidence that the change in an object's motion depends on the sum of the forces on the object and the mass of the object. **Clarification Statement:** Emphasis is on balanced (Newton's first law) and unbalanced forces in a system, qualitative comparisons of forces, mass and changes in motion (Newton's second law), frame of reference, and specification of units. **Assessment Boundary:** Assessment is limited to forces and changes in motion in one dimension in an inertial reference frame and to change in one variable at a time. Assessment does not include the use of trigonometry.	**PS2.A. Forces and Motion** The motion of an object is determined by the sum of the forces acting on it; if the total force on the object is not zero, its motion will change. The greater the mass of the object, the greater the force needed to achieve the same change in motion. For any given object, a larger force causes a larger change in motion. All positions of objects and the directions of forces and motions must be described in an arbitrarily chosen reference frame and arbitrarily chosen units of size. In order to share information with other people, these choices must also be shared.
MS-PS2-3. Ask questions about data to determine the factors that affect the strength of electric and magnetic forces. **Clarification Statement:** Examples of devices that use electric and magnetic forces could include electromagnets, electric motors, or generators. Examples of data could include the effect of the number of turns of wire on the strength of an electromagnet, or the effect of increasing the number or strength of magnets on the speed of an electric motor. **Assessment Boundary:** Assessment about questions that require quantitative answers is limited to proportional reasoning and algebraic thinking.	**PS2.B. Types of Interactions** Electric and magnetic (electromagnetic) forces can be attractive or repulsive, and their sizes depend on the magnitudes of the charges, currents, or magnetic strengths involved and on the distances between the interacting objects.

Performance Expectations and Disciplinary Core Ideas for Physical Science (*continued*)

Performance Expectations (PEs)	Disciplinary Core Ideas (DCIs)
MS-PS2-4. Construct and present arguments using evidence to support the claim that gravitational interactions are attractive and depend on the masses of interacting objects. **Clarification Statement:** Examples of evidence for arguments could include data generated from simulations or digital tools; and charts displaying mass, strength of interaction, distance from the Sun, and orbital periods of objects within the solar system. **Assessment Boundary:** Assessment does not include Newton's law of gravitation or Kepler's laws.	**PS2.B. Types of Interactions** Gravitational forces are always attractive. There is a gravitational force between any two masses, but it is very small except when one or both of the objects have large mass—e.g., Earth and the Sun.
MS-PS2-5. Conduct an investigation and evaluate the experimental design to provide evidence that fields exist between objects exerting forces on each other even though the objects are not in contact. **Clarification Statement:** Examples of this phenomenon could include the interactions of magnets, electrically charged strips of tape, and electrically charged pith balls. Examples of investigations could include firsthand experiences or simulations. **Assessment Boundary:** Assessment is limited to electric and magnetic fields, and limited to qualitative evidence for the existence of fields.	**PS2.B. Types of Interactions** Forces that act at a distance (electric, magnetic, and gravitational) can be explained by fields that extend through space and can be mapped by their effect on a test object (a charged object, or a ball, respectively).
MS-PS3-1. Construct and interpret graphical displays of data to describe the relationships of kinetic energy to the mass of an object and to the speed of an object. **Clarification Statement:** Emphasis is on descriptive relationships between kinetic energy and mass separately from kinetic energy and speed. Examples could include riding a bicycle at different speeds, rolling different sizes of rocks downhill, and getting hit by a wiffle ball versus a tennis ball.	**PS3.A. Definitions of Energy** Motion energy is properly called kinetic energy; it is proportional to the mass of the moving object and grows with the square of its speed.
MS-PS3-2. Develop a model to describe that when the arrangement of objects interacting at a distance changes, different amounts of potential energy are stored in the system. **Clarification Statement:** Emphasis is on relative amounts of potential energy, not on calculations of potential energy. Examples of objects within systems interacting at varying distances could include the Earth and either a roller coaster cart at varying positions on a hill or objects at varying heights on shelves, changing the direction/orientation of a magnet, and a balloon with static electrical charge being brought closer to a classmate's hair. Examples of models could include representations, diagrams, pictures, and written descriptions of systems. **Assessment Boundary:** Assessment is limited to two objects and electric, magnetic, and gravitational interactions.	**PS3.A. Definitions of Energy** A system of objects may also contain stored (potential) energy, depending on their relative positions. **PS3.C. Relationship Between Energy and Forces** When two objects interact, each one exerts a force on the other that can cause energy to be transferred to or from the object.

Performance Expectations and Disciplinary Core Ideas for Physical Science (*continued*)

Performance Expectations (PEs)	Disciplinary Core Ideas (DCIs)
MS-PS3-3. Apply scientific principles to design, construct, and test a device that either minimizes or maximizes thermal energy transfer. **Clarification Statement:** Examples of devices could include an insulated box, a solar cooker, and a Styrofoam cup. **Assessment Boundary:** Assessment does not include calculating the total amount of thermal energy transferred.	**PS3.A. Definitions of Energy** Temperature is not a measure of energy; the relationship between the temperature and the total energy of a system depends on the types, states, and amounts of matter present. (MS-PS3-4) **PS3.B. Conservation of Energy and Energy Transfer** Energy is spontaneously transferred out of hotter regions or objects and into colder ones. **ETS1.A. Defining and Delimiting Engineering Problems** The more precisely a design task's criteria and constraints can be defined, the more likely it is that the designed solution will be successful. Specification of constraints includes consideration of scientific principles and other relevant knowledge that is likely to limit possible solutions. (MS-ETS1-1) **ETS1.B. Developing Possible Solutions** A solution needs to be tested, and then modified on the basis of the test results in order to improve it. There are systematic processes for evaluating solutions with respect to how well they meet criteria and constraints of a problem. (MS-ETS1-4), (MS-PS1-6)
MS-PS3-4. Plan an investigation to determine the relationships among the energy transferred, the type of matter, the mass, and the change in the average kinetic energy of the particles as measured by the temperature of the sample. **Clarification Statement:** Examples of experiments could include comparing final water temperatures after different masses of ice melted in the same volume of water with the same initial temperature, the temperature change of samples of different materials with the same mass as they cool or heat in the environment, or the same material with different masses when a specific amount of energy is added. **Assessment Boundary:** Assessment does not include calculating the total amount of thermal energy transferred.	**PS3.A. Definitions of Energy** Temperature is not a measure of energy; the relationship between the temperature and the total energy of a system depends on the types, states, and amounts of matter present. (MS-PS3-3) **PS3.B. Conservation of Energy and Energy Transfer** The amount of energy transfer needed to change the temperature of a matter sample by a given amount depends on the nature of the matter, the size of the sample, and the environment.
MS-PS3-5. Construct, use, and present arguments to support the claim that when the kinetic energy of an object changes, energy is transferred to or from the object. **Clarification Statement:** Examples of empirical evidence used in arguments could include an inventory or other representation of the energy before and after the transfer in the form of temperature changes or motion of object. **Assessment Boundary:** Assessment does not include calculations of energy.	**PS3.B. Conservation of Energy and Energy Transfer** When the motion energy of an object changes, there is inevitably some other change in energy at the same time.

Performance Expectations and Disciplinary Core Ideas for Physical Science (*continued*)

Performance Expectations (PEs)	Disciplinary Core Ideas (DCIs)
MS-PS4-1. Use mathematical representations to describe a simple model for waves that includes how the amplitude of a wave is related to the energy in a wave. **Clarification Statement:** Emphasis is on describing waves with both qualitative and quantitative thinking. **Assessment Boundary:** Assessment does not include electromagnetic waves and is limited to standard repeating waves.	**PS4.A. Wave Properties** A simple wave has a repeating pattern with a specific wavelength, frequency, and amplitude.
MS-PS4-2. Develop and use a model to describe that waves are reflected, absorbed, or transmitted through various materials. **Clarification Statement:** Emphasis is on both light and mechanical waves. Examples of models could include drawings, simulations, and written descriptions. **Assessment Boundary:** Assessment is limited to qualitative applications pertaining to light and mechanical waves.	**PS4.A. Wave Properties** A sound wave needs a medium through which it is transmitted. **PS4.B. Electromagnetic Radiation** When light shines on an object, it is reflected, absorbed, or transmitted through the object, depending on the object's material and the frequency (color) of the light. The path that light travels can be traced as straight lines, except at surfaces between different transparent materials (e.g., air and water, air and glass) where the light path bends. A wave model of light is useful for explaining brightness, color, and the frequency-dependent bending of light at a surface between media. However, because light can travel through space, it cannot be a matter wave, like sound or water waves.
MS-PS4-3. Integrate qualitative scientific and technical information to support the claim that digitized signals are a more reliable way to encode and transmit information than analog signals. **Clarification Statement:** Emphasis is on a basic understanding that waves can be used for communication purposes. Examples could include using fiber-optic cable to transmit light pulses, radio wave pulses in Wi-Fi devices, and conversion of stored binary patterns to make sound or text on a computer screen. **Assessment Boundary:** Assessment does not include binary counting. Assessment does not include the specific mechanism of any given device.	**PS4.C. Information Technologies and Instrumentation** Digitized signals (sent as wave pulses) are a more reliable way to encode and transmit information.

Performance Expectations and Disciplinary Core Ideas for Life Science

Performance Expectations (PEs)	Disciplinary Core Ideas (DCIs)
MS-LS1-1. Conduct an investigation to provide evidence that living things are made of cells; either one cell or many different numbers and types of cells. **Clarification Statement:** Emphasis is on developing evidence that living things are made of cells, distinguishing between living and nonliving things, and understanding that living things may be made of one cell or many and varied cells.	**LS1.A. Structure and Function** All living things are made up of cells, which is the smallest unit that can be said to be alive. An organism may consist of one single cell (unicellular) or many different numbers and types of cells (multicellular).
MS-LS1-2. Develop and use a model to describe the function of a cell as a whole and ways that parts of cells contribute to the function. **Clarification Statement:** Emphasis is on the cell functioning as a whole system and the primary role of identified parts of the cell, specifically the nucleus, chloroplasts, mitochondria, cell membrane, and cell wall. **Assessment Boundary:** Assessment of organelle structure/function relationships is limited to the cell wall and cell membrane. Assessment of the function of the other organelles is limited to their relationship to the whole cell. Assessment does not include the biochemical function of cells or cell parts.	**LS1.A. Structure and Function** Within cells, special structures are responsible for particular functions, and the cell membrane forms the boundary that controls what enters and leaves the cell.
MS-LS1-3. Use argument supported by evidence for how the body is a system of interacting subsystems composed of groups of cells. **Clarification Statement:** Emphasis is on the conceptual understanding that cells form tissues and tissues form organs specialized for particular body functions. Examples could include the interaction of subsystems within a system and the normal functioning of those systems. **Assessment Boundary:** Assessment does not include the mechanism of one body system independent of others. Assessment is limited to the circulatory, excretory, digestive, respiratory, muscular, and nervous systems.	**LS1.A. Structure and Function** In multicellular organisms, the body is a system of multiple interacting subsystems. These subsystems are groups of cells that work together to form tissues and organs that are specialized for particular body functions.
MS-LS1-4. Use argument based on empirical evidence and scientific reasoning to support an explanation for how characteristic animal behaviors and specialized plant structures affect the probability of successful reproduction of animals and plants respectively. **Clarification Statement:** Examples of behaviors that affect the probability of animal reproduction could include nest building to protect young from cold, herding of animals to protect young from predators, and vocalization of animals and colorful plumage to attract mates for breeding. Examples of animal behaviors that affect the probability of plant reproduction could include transferring pollen or seeds and creating conditions for seed germination and growth. Examples of plant structures could include bright flowers attracting butterflies that transfer pollen, flower nectar and odors that attract insects that transfer pollen, and hard shells on nuts that squirrels bury.	**LS1.B. Growth and Development of Organisms** Animals engage in characteristic behaviors that increase the odds of reproduction. Plants reproduce in a variety of ways, sometimes depending on animal behavior and specialized features for reproduction.

Performance Expectations and Disciplinary Core Ideas for Life Science (*continued*)

Performance Expectations (PEs)	Disciplinary Core Ideas (DCIs)
MS-LS1-5. Construct a scientific explanation based on evidence for how environmental and genetic factors influence the growth of organisms. **Clarification Statement:** Examples of local environmental conditions could include availability of food, light, space, and water. Examples of genetic factors could include large-breed cattle and species of grass affecting growth of organisms. Examples of evidence could include drought decreasing plant growth, fertilizer increasing plant growth, different varieties of plant seeds growing at different rates in different conditions, and fish growing larger in large ponds than they do in small ponds. **Assessment Boundary:** Assessment does not include genetic mechanisms, gene regulation, or biochemical processes.	**LS1.B. Growth and Development of Organisms** Genetic factors as well as local conditions affect the growth of the adult plant.
MS-LS1-6. Construct a scientific explanation based on evidence for the role of photosynthesis in the cycling of matter and flow of energy into and out of organisms. **Clarification Statement:** Emphasis is on tracing movement of matter and flow of energy. **Assessment Boundary:** Assessment does not include the biochemical mechanisms of photosynthesis.	**PS3.D. Energy in Chemical Processes and Everyday Life** The chemical reaction by which plants produce complex food molecules (sugars) requires an energy input (i.e., from sunlight) to occur. In this reaction, carbon dioxide and water combine to form carbon-based organic molecules and release oxygen. **LS1.C. Organization for Matter and Energy Flow in Organisms** Plants, algae (including phytoplankton), and many microorganisms use the energy from light to make sugars (food) from carbon dioxide from the atmosphere and water through the process of photosynthesis, which also releases oxygen. These sugars can be used immediately or stored for growth or later use.
MS-LS1-7. Develop a model to describe how food is rearranged through chemical reactions forming new molecules that support growth and/or release energy as this matter moves through an organism. **Clarification Statement:** Emphasis is on describing that molecules are broken apart and put back together and that in this process, energy is released. **Assessment Boundary:** Assessment does not include details of the chemical reactions for photosynthesis or respiration.	**LS1.C. Organization for Matter and Energy Flow in Organisms** Within individual organisms, food moves through a series of chemical reactions in which it is broken down and rearranged to form new molecules, to support growth, or to release energy. **PS3.D. Energy in Chemical Processes and Everyday Life** Cellular respiration in plants and animals involves chemical reactions with oxygen that release stored energy. In these processes, complex molecules containing carbon react with oxygen to produce carbon dioxide and other materials.
MS-LS1-8. Gather and synthesize information that sensory receptors respond to stimuli by sending messages to the brain for immediate behavior or storage as memories. **Assessment Boundary:** Assessment does not include mechanisms for the transmission of this information.	**LS1.D. Information Processing** Each sense receptor responds to different inputs (electromagnetic, mechanical, chemical), transmitting them as signals that travel along nerve cells to the brain. The signals are then processed in the brain, resulting in immediate behaviors or memories.

Performance Expectations and Disciplinary Core Ideas for Life Science (*continued*)

Performance Expectations (PEs)	Disciplinary Core Ideas (DCIs)
MS-LS2-1. Analyze and interpret data to provide evidence for the effects of resource availability on organisms and populations of organisms in an ecosystem. **Clarification Statement:** Emphasis is on cause-and-effect relationships between resources and growth of individual organisms and the numbers of organisms in ecosystems during periods of abundant and scarce resources.	**LS2.A. Interdependent Relationships in Ecosystems** Organisms, and populations of organisms, are dependent on their environmental interactions both with other living things and with nonliving factors. In any ecosystem, organisms and populations with similar requirements for food, water, oxygen, or other resources may compete with each other for limited resources, access to which consequently constrains their growth and reproduction. Growth of organisms and population increases are limited by access to resources.
MS-LS2-2. Construct an explanation that predicts patterns of interactions among organisms across multiple ecosystems. **Clarification Statement:** Emphasis is on predicting consistent patterns of interactions in different ecosystems in terms of the relationships among and between organisms and abiotic components of ecosystems. Examples of types of interactions could include competitive, predatory, and mutually beneficial.	**LS2.A. Interdependent Relationships in Ecosystems** Similarly, predatory interactions may reduce the number of organisms or eliminate whole populations of organisms. Mutually beneficial interactions, in contrast, may become so interdependent that each organism requires the other for survival. Although the species involved in these competitive, predatory, and mutually beneficial interactions vary across ecosystems, the patterns of interactions of organisms with their environments, both living and nonliving, are shared.
MS-LS2-3. Develop a model to describe the cycling of matter and flow of energy among living and nonliving parts of an ecosystem. **Clarification Statement:** Emphasis is on describing the conservation of matter and flow of energy into and out of various ecosystems, and on defining the boundaries of the system. **Assessment Boundary:** Assessment does not include the use of chemical reactions to describe the processes.	**LS2.B. Cycles of Matter and Energy Transfer in Ecosystems** Food webs are models that demonstrate how matter and energy is transferred between producers, consumers, and decomposers as the three groups interact within an ecosystem. Transfers of matter into and out of the physical environment occur at every level. Decomposers recycle nutrients from dead plant or animal matter back to the soil in terrestrial environments or to the water in aquatic environments. The atoms that make up the organisms in an ecosystem are cycled repeatedly between the living and nonliving parts of the ecosystem.
MS-LS2-4. Construct an argument supported by empirical evidence that changes to physical or biological components of an ecosystem affect populations. **Clarification Statement:** Emphasis is on recognizing patterns in data and making warranted inferences about changes in populations, and on evaluating empirical evidence supporting arguments about changes to ecosystems.	**LS2.C. Ecosystem Dynamics, Functioning, and Resilience** Ecosystems are dynamic in nature; their characteristics can vary over time. Disruptions to any physical or biological component of an ecosystem can lead to shifts in all its populations.
MS-LS2-5. Evaluate competing design solutions for maintaining biodiversity and ecosystem services. **Clarification Statement:** Examples of ecosystem services could include water purification, nutrient recycling, and prevention of soil erosion. Examples of design solution constraints could include scientific, economic, and social considerations.	**LS2.C. Ecosystem Dynamics, Functioning, and Resilience** Biodiversity describes the variety of species found in Earth's terrestrial and oceanic ecosystems. The completeness or integrity of an ecosystem's biodiversity is often used as a measure of its health. **LS4.D. Biodiversity and Humans** Changes in biodiversity can influence humans' resources, such as food, energy, and medicines, as well as ecosystem services that humans rely on—for example, water purification and recycling. **ETS1.B. Developing Possible Solutions** There are systematic processes for evaluating solutions with respect to how well they meet the criteria and constraints of a problem. (MS-ETS1-2), (MS-ETS1-3)

Performance Expectations and Disciplinary Core Ideas for Life Science (*continued*)

Performance Expectations (PEs)	Disciplinary Core Ideas (DCIs)
MS-LS3-1. Develop and use a model to describe why structural changes to genes (mutations) located on chromosomes may affect proteins and may result in harmful, beneficial, or neutral effects to the structure and function of the organism. **Clarification Statement:** Emphasis is on conceptual understanding that changes in genetic material may result in making different proteins. **Assessment Boundary:** Assessment does not include specific changes at the molecular level, mechanisms for protein synthesis, or specific types of mutations.	**LS3.A. Inheritance of Traits** Genes are located in the chromosomes of cells, with each chromosome pair containing two variants of each of many distinct genes. Each distinct gene chiefly controls the production of specific proteins, which in turn affects the traits of the individual. Changes (mutations) to genes can result in changes to proteins, which can affect the structures and functions of the organism and thereby change traits. **LS3.B. Variation of Traits** In addition to variations that arise from sexual reproduction, genetic information can be altered because of mutations. Though rare, mutations may result in changes to the structure and function of proteins. Some changes are beneficial, others harmful, and some neutral to the organism.
MS-LS3-2. Develop and use a model to describe why asexual reproduction results in offspring with identical genetic information and sexual reproduction results in offspring with genetic variation. **Clarification Statement:** Emphasis is on using models such as Punnett squares, diagrams, and simulations to describe the cause-and-effect relationship of gene transmission from parent(s) to offspring and resulting genetic variation.	**LS1.B. Growth and Development of Organisms** Organisms reproduce, either sexually or asexually, and transfer their genetic information to their offspring. **LS3.A. Inheritance of Traits** Variations of inherited traits between parent and offspring arise from genetic differences that result from the subset of chromosomes (and therefore genes) inherited. **LS3.B. Variation of Traits** In sexually reproducing organisms, each parent contributes half of the genes acquired (at random) by the offspring. Individuals have two of each chromosome and hence two alleles of each gene, one acquired from each parent. These versions may be identical or may differ from each other.
MS-LS4-1. Analyze and interpret data for patterns in the fossil record that document the existence, diversity, extinction, and change of life forms throughout the history of life on Earth under the assumption that natural laws operate today as in the past. **Clarification Statement:** Emphasis is on finding patterns of changes in the level of complexity of anatomical structures in organisms and the chronological order of fossil appearance in the rock layers. **Assessment Boundary:** Assessment does not include the names of individual species or geologic eras in the fossil record.	**LS4.A. Evidence of Common Ancestry and Diversity** The collection of fossils and their placement in chronological order (e.g., through the location of the sedimentary layers in which they are found or through radioactive dating) is known as the fossil record. It documents the existence, diversity, extinction, and change of many life forms throughout the history of life on Earth.
MS-LS4-2. Apply scientific ideas to construct an explanation for the anatomical similarities and differences among modern organisms and between modern and fossil organisms to infer evolutionary relationships. **Clarification Statement:** Emphasis is on explanations of the evolutionary relationships among organisms in terms of similarity or differences of the gross appearance of anatomical structures.	**LS4.A. Evidence of Common Ancestry and Diversity** Anatomical similarities and differences between various organisms living today and between them and organisms in the fossil record, enable the reconstruction of evolutionary history and the inference of lines of evolutionary descent.

Performance Expectations and Disciplinary Core Ideas for Life Science (*continued*)

Performance Expectations (PEs)	Disciplinary Core Ideas (DCIs)
MS-LS4-3. Analyze displays of pictorial data to compare patterns of similarities in the embryological development across multiple species to identify relationships not evident in the fully formed anatomy. **Clarification Statement:** Emphasis is on inferring general patterns of relatedness among embryos of different organisms by comparing the macroscopic appearance of diagrams or pictures. **Assessment Boundary:** Assessment of comparisons is limited to gross appearance of anatomical structures in embryological development.	**LS4.A. Evidence of Common Ancestry and Diversity** Comparison of the embryological development of different species also reveals similarities that show relationships not evident in the fully formed anatomy.
MS-LS4-4. Construct an explanation based on evidence that describes how genetic variations of traits in a population increase some individuals' probability of surviving and reproducing in a specific environment. **Clarification Statement:** Emphasis is on using simple probability statements and proportional reasoning to construct explanations.	**LS4.B. Natural Selection** Natural selection leads to the predominance of certain traits in a population, and the suppression of others.
MS-LS4-5. Gather and synthesize information about the technologies that have changed the way humans influence the inheritance of desired traits in organisms. **Clarification Statement:** Emphasis is on synthesizing information from reliable sources about the influence of humans on genetic outcomes in artificial selection (such as genetic modification, animal husbandry, gene therapy) and on the impacts these technologies have on society as well as the technologies leading to these scientific discoveries.	**LS4.B. Natural Selection** In artificial selection, humans have the capacity to influence certain characteristics of organisms by selective breeding. One can choose desired parental traits determined by genes, which are then passed on to offspring.
MS-LS4-6. Use mathematical representations to support explanations of how natural selection may lead to increases and decreases of specific traits in populations over time. **Clarification Statement:** Emphasis is on using mathematical models, probability statements, and proportional reasoning to support explanations of trends in changes to populations over time. **Assessment Boundary:** Assessment does not include Hardy-Weinberg calculations.	**LS4.C. Adaptation** Adaptation by natural selection acting over generations is one important process by which species change over time in response to changes in environmental conditions. Traits that support successful survival and reproduction in the new environment become more common; those that do not become less common. Thus, the distribution of traits in a population changes.

Performance Expectations and Disciplinary Core Ideas for Earth and Space Science

Performance Expectations (PEs)	Disciplinary Core Ideas (DCIs)
MS-ESS1-1. Develop and use a model of the Earth-Sun-Moon system to describe the cyclic patterns of lunar phases, eclipses of the Sun and Moon, and seasons. **Clarification Statement:** Examples of models can be physical, graphical, or conceptual.	**ESS1.A. The Universe and Its Stars** Patterns of the apparent motion of the Sun, the Moon, and stars in the sky can be observed, described, predicted, and explained with models. **ESS1.B. Earth and the Solar System** This model of the solar system can explain eclipses of the Sun and the Moon. Earth's spin axis is fixed in direction over the short term but tilted relative to its orbit around the Sun. The seasons are a result of that tilt and are caused by the differential intensity of sunlight on different areas of Earth across the year.
MS-ESS1-2. Develop and use a model to describe the role of gravity in the motions within galaxies and the solar system. **Clarification Statement:** Emphasis for the model is on gravity as the force that holds together the solar system and Milky Way galaxy and controls orbital motions within them. Examples of models can be physical (such as the analogy of distance along a football field or computer visualizations of elliptical orbits) or conceptual (such as mathematical proportions relative to the size of familiar objects such as students' school or state). **Assessment Boundary:** Assessment does not include Kepler's laws of orbital motion or the apparent retrograde motion of the planets as viewed from Earth.	**ESS1.A. The Universe and Its Stars** Earth and its solar system are part of the Milky Way galaxy, which is one of many galaxies in the universe. **ESS1.B. Earth and the Solar System** The solar system consists of the Sun and a collection of objects, including planets, their moons, and asteroids that are held in orbit around the Sun by its gravitational pull on them. (MS-ESS1-3) The solar system appears to have formed from a disk of dust and gas, drawn together by gravity.
MS-ESS1-3. Analyze and interpret data to determine scale properties of objects in the solar system. **Clarification Statement:** Emphasis is on the analysis of data from Earth-based instruments, space-based telescopes, and spacecraft to determine similarities and differences among solar system objects. Examples of scale properties include the sizes of an object's layers (such as crust and atmosphere), surface features (such as volcanoes), and orbital radius. Examples of data include statistical information, drawings and photographs, and models. **Assessment Boundary:** Assessment does not include recalling facts about properties of the planets and other solar system bodies.	**ESS1.B. Earth and the Solar System** The solar system consists of the Sun and a collection of objects, including planets, their moons, and asteroids that are held in orbit around the Sun by its gravitational pull on them. (MS-ESS1-2)

Performance Expectations and Disciplinary Core Ideas for Earth and Space Science (*continued*)

Performance Expectations (PEs)	Disciplinary Core Ideas (DCIs)
MS-ESS1-4. Construct a scientific explanation based on evidence from rock strata for how the geologic time scale is used to organize Earth's 4.6-billion-year-old history. **Clarification Statement:** Emphasis is on how analyses of rock formations and the fossils they contain are used to establish relative ages of major events in Earth's history. Examples of Earth's major events could range from being very recent (such as the last Ice Age or the earliest fossils of *Homo sapiens*) to very old (such as the formation of Earth or the earliest evidence of life). Examples can include the formation of mountain chains and ocean basins, the evolution or extinction of particular living organisms, or significant volcanic eruptions. **Assessment Boundary:** Assessment does not include recalling the names of specific periods or epochs and events within them.	**ESS1.C. The History of Planet Earth** The geologic time scale interpreted from rock strata provides a way to organize Earth's history. Analyses of rock strata and the fossil record provide only relative dates, not an absolute scale.
MS-ESS2-1. Develop a model to describe the cycling of Earth's materials and the flow of energy that drives this process. **Clarification Statement:** Emphasis is on the processes of melting, crystallization, weathering, deformation, and sedimentation, which act together to form minerals and rocks through the cycling of Earth's materials. **Assessment Boundary:** Assessment does not include the identification and naming of minerals.	**ESS2.A. Earth Materials and Systems** All Earth processes are the result of energy flowing and matter cycling within and among the planet's systems. This energy is derived from the Sun and Earth's hot interior. The energy that flows and matter that cycles produce chemical and physical changes in Earth's materials and living organisms.
MS-ESS2-2. Construct an explanation based on evidence for how geoscience processes have changed Earth's surface at varying time and spatial scales. **Clarification Statement:** Emphasis is on how processes change Earth's surface at time and spatial scales that can be large (such as slow plate motions or the uplift of large mountain ranges) or small (such as rapid landslides or microscopic geochemical reactions), and how many geoscience processes (such as earthquakes, volcanoes, and meteor impacts) usually behave gradually but are punctuated by catastrophic events. Examples of geoscience processes include surface weathering and deposition by the movements of water, ice, and wind. Emphasis is on geoscience processes that shape local geographic features, where appropriate.	**ESS2.A. Earth Materials and Systems** The planet's systems interact over scales that range from microscopic to global in size, and they operate over fractions of a second to billions of years. These interactions have shaped Earth's history and will determine its future. **ESS2.C. The Roles of Water in Earth's Surface Processes** Water's movements—both on the land and underground—cause weathering and erosion, which change the land's surface features and create underground formations.
MS-ESS2-3. Analyze and interpret data on the distribution of fossils and rocks, continental shapes, and seafloor structures to provide evidence of the past plate motions. **Clarification Statement:** Examples of data include similarities of rock and fossil types on different continents, the shapes of the continents (including continental shelves), and the locations of ocean structures (such as ridges, fracture zones, and trenches). **Assessment Boundary:** Paleomagnetic anomalies in oceanic and continental crust are not assessed.	**ESS2.B. Plate Tectonics and Large-Scale System Interactions** Maps of ancient land and water patterns, based on investigations of rocks and fossils, make clear how Earth's plates have moved great distances, collided, and spread apart. **ESS1.C. The History of Planet Earth** Tectonic processes continually generate new ocean sea floor at ridges and destroy old seafloor at trenches.

Performance Expectations and Disciplinary Core Ideas for Earth and Space Science (*continued*)

Performance Expectations (PEs)	Disciplinary Core Ideas (DCIs)
MS-ESS2-4. Develop a model to describe the cycling of water through Earth's systems driven by energy from the Sun and the force of gravity. **Clarification Statement:** Emphasis is on the ways that water changes its state as it moves through the multiple pathways of the hydrologic cycle. Examples of models can be conceptual or physical. **Assessment Boundary:** A quantitative understanding of the latent heats of vaporization and fusion is not assessed.	**ESS2.C. The Roles of Water in Earth's Surface Processes** Water continually cycles among land, ocean, and atmosphere via transpiration, evaporation, condensation and crystallization, and precipitation, as well as downhill flows on land. **ESS2.C. The Roles of Water in Earth's Surface Processes** Global movements of water and its changes in form are propelled by sunlight and gravity.
MS-ESS2-5. Collect data to provide evidence for how the motions and complex interactions of air masses results in changes in weather conditions. **Clarification Statement:** Emphasis is on how air masses flow from regions of high pressure to low pressure, causing weather (defined by temperature, pressure, humidity, precipitation, and wind) at a fixed location to change over time, and how sudden changes in weather can result when different air masses collide. Emphasis is on how weather can be predicted within probabilistic ranges. Examples of data can be provided to students (such as weather maps, diagrams, and visualizations) or obtained through laboratory experiments (such as with condensation). **Assessment Boundary:** Assessment does not include recalling the names of cloud types or weather symbols used on weather maps or the reported diagrams from weather stations.	**ESS2.C. The Roles of Water in Earth's Surface Processes** The complex patterns of the changes and the movement of water in the atmosphere, determined by winds, landforms, and ocean temperatures and currents, are major determinants of local weather patterns. **ESS2.D. Weather and Climate** Because these patterns are so complex, weather can only be predicted probabilistically.
MS-ESS2-6. Develop and use a model to describe how unequal heating and rotation of the Earth cause patterns of atmospheric and oceanic circulation that determine regional climates. **Clarification Statement:** Emphasis is on how patterns vary by latitude, altitude, and geographic land distribution. Emphasis of atmospheric circulation is on the sunlight-driven latitudinal banding, the Coriolis effect, and resulting prevailing winds; emphasis of ocean circulation is on the transfer of heat by the global ocean convection cycle, which is constrained by the Coriolis effect and the outlines of continents. Examples of models can be diagrams, maps and globes, or digital representations. **Assessment Boundary:** Assessment does not include the dynamics of the Coriolis effect.	**ESS2.C. The Roles of Water in Earth's Surface Processes** Variations in density due to variations in temperature and salinity drive a global pattern of interconnected ocean currents. **ESS2.D. Weather and Climate** The ocean exerts a major influence on weather and climate by absorbing energy from the Sun, releasing it over time, and globally redistributing it through ocean currents. **ESS2.D. Weather and Climate** Weather and climate are influenced by interactions involving sunlight, the ocean, the atmosphere, ice, landforms, and living things. These interactions vary with latitude, altitude, and local and regional geography, all of which can affect oceanic and atmospheric flow patterns.

Performance Expectations and Disciplinary Core Ideas for Earth and Space Science (*continued*)

Performance Expectations (PEs)	Disciplinary Core Ideas (DCIs)
MS-ESS3-1. Construct a scientific explanation based on evidence for how the uneven distributions of Earth's mineral, energy, and groundwater resources are the result of past and current geoscience processes. **Clarification Statement:** Emphasis is on how these resources are limited and typically non-renewable, and how their distributions are significantly changing as a result of removal by humans. Examples of uneven distributions of resources as a result of past processes include but are not limited to petroleum (locations of the burial of organic marine sediments and subsequent geologic traps), metal ores (locations of past volcanic and hydrothermal activity associated with subduction zones), and soil (locations of active weathering and/or deposition of rock).	**ESS3.A. Natural Resources** Humans depend on Earth's land, ocean, atmosphere, and biosphere for many different resources. Minerals, fresh water, and biosphere resources are limited, and many are not renewable or replaceable over human lifetimes. These resources are distributed unevenly around the planet as a result of past geologic processes.
MS-ESS3-2. Analyze and interpret data on natural hazards to forecast future catastrophic events and inform the development of technologies to mitigate their effects. **Clarification Statement:** Emphasis is on how some natural hazards, such as volcanic eruptions and severe weather, are preceded by phenomena that allow for reliable predictions, but others, such as earthquakes, occur suddenly and with no notice, and thus are not yet predictable. Examples of natural hazards can be taken from interior processes (such as earthquakes and volcanic eruptions), surface processes (such as mass wasting and tsunamis), or severe weather events (such as hurricanes, tornadoes, and floods). Examples of data can include the locations, magnitudes, and frequencies of the natural hazards. Examples of technologies can be global (such as satellite systems to monitor hurricanes or forest fires) or local (such as building basements in tornado-prone regions or reservoirs to mitigate droughts).	**ESS3.B. Natural Hazards** Mapping the history of natural hazards in a region, combined with an understanding of related geologic forces, can help forecast the locations and likelihoods of future events.
MS-ESS3-3. Apply scientific principles to design a method for monitoring and minimizing a human impact on the environment. **Clarification Statement:** Examples of the design process include examining human environmental impacts, assessing the kinds of solutions that are feasible, and designing and evaluating solutions that could reduce that impact. Examples of human impacts can include water usage (such as the withdrawal of water from streams and aquifers or the construction of dams and levees), land usage (such as urban development, agriculture, or the removal of wetlands), and pollution (such as of the air, water, or land).	**ESS3.C. Human Impacts on Earth Systems** Human activities have significantly altered the biosphere, sometimes damaging or destroying natural habitats and causing the extinction of other species. But changes to Earth's environments can have different impacts (negative and positive) for different living things. Typically as human populations and per capita consumption of natural resources increase, so do the negative impacts on Earth unless the activities and technologies involved are engineered otherwise. (MS-ESS3-4)

Performance Expectations and Disciplinary Core Ideas for Earth and Space Science (*continued*)

Performance Expectations (PEs)	Disciplinary Core Ideas (DCIs)
MS-ESS3-4. Construct an argument supported by evidence for how increases in human population and per capita consumption of natural resources impact Earth's systems. **Clarification Statement:** Examples of evidence include grade-appropriate databases on human populations and the rates of consumption of food and natural resources (such as freshwater, mineral, and energy). Examples of impacts can include changes to the appearance, composition, and structure of Earth's systems as well as the rates at which they change. The consequences of increases in human populations and consumption of natural resources are described by science, but science does not make the decisions for the actions society takes.	**ESS3.C. Human Impacts on Earth Systems** Typically as human populations and per capita consumption of natural resources increase, so do the negative impacts on Earth unless the activities and technologies involved are engineered otherwise. (MS-ESS3-3)
MS-ESS3-5. Ask questions to clarify evidence of the factors that have caused the rise in global temperatures over the past century. **Clarification Statement:** Examples of factors include human activities (such as fossil fuel combustion, cement production, and agricultural activity) and natural processes (such as changes in incoming solar radiation or volcanic activity). Examples of evidence can include tables, graphs, and maps of global and regional temperatures, atmospheric levels of gases such as carbon dioxide and methane, and the rates of human activities. Emphasis is on the major role that human activities play in causing the rise in global temperatures.	**ESS3.D. Global Climate Change** Human activities, such as the release of greenhouse gases from burning fossil fuels, are major factors in the current rise in Earth's mean surface temperature (global warming). Reducing the level of climate change and reducing human vulnerability to whatever climate changes do occur depend on the understanding of climate science, engineering capabilities, and other kinds of knowledge, such as understanding of human behavior, and on applying that knowledge wisely in decisions and activities.

Performance Expectations and Disciplinary Core Ideas for Engineering Design

Performance Expectations (PEs)	Disciplinary Core Ideas (DCIs)
MS-ETS1-1. Define the criteria and constraints of a design problem with sufficient precision to ensure a successful solution, taking into account relevant scientific principles and potential impacts on people and the natural environment that may limit possible solutions.	**ETS1.A. Defining and Delimiting Engineering Problems** The more precisely a design task's criteria and constraints can be defined, the more likely it is that the designed solution will be successful. Specification of constraints includes consideration of scientific principles and other relevant knowledge that is likely to limit possible solutions. (MS-PS3-3)
MS-ETS1-2. Evaluate competing design solutions using a systematic process to determine how well they meet the criteria and constraints of the problem.	**ETS1.B. Developing Possible Solutions** There are systematic processes for evaluating solutions with respect to how well they meet the criteria and constraints of a problem. (MS-ETS1-3), (MS-LS2-5)
MS-ETS1-3. Analyze data from tests to determine similarities and differences among several design solutions to identify the best characteristics of each that can be combined into a new solution to better meet the criteria for success.	**ETS1.B. Developing Possible Solutions** There are systematic processes for evaluating solutions with respect to how well they meet the criteria and constraints of a problem. (MS-ETS1-2), (MS-LS2-5) Sometimes parts of different solutions can be combined to create a solution that is better than any of its predecessors. **ETS1.C. Optimizing the Design Solution** Although one design may not perform the best across all tests, identifying the characteristics of the design that performed the best in each test can provide useful information for the redesign process—that is, some of the characteristics may be incorporated into the new design.
MS-ETS1-4. Develop a model to generate data for iterative testing and modification of a proposed object, tool, or process such that an optimal design can be achieved.	**ETS1.B. Developing Possible Solutions** A solution needs to be tested, and then modified on the basis of the test results in order to improve it. (MS-PS1-6), (MS-PS3-3) Models of all kinds are important for testing solutions. **ETS1.C. Optimizing the Design Solution** The iterative process of testing the most promising solutions and modifying what is proposed on the basis of the test results leads to greater refinement and ultimately to an optimal solution. (MS-PS1-6)

CHAPTER 6
Focus on High School

Science and Engineering Practices

Asking Questions and Defining Problems for Grades 9–12

Asking questions and defining problems in 9–12 builds on K–8 experiences and progresses to formulating, refining, and evaluating empirically testable questions and design problems using models and simulations.

- Ask questions that arise from careful observation of phenomena, or unexpected results, to clarify and/or seek additional information.
- Ask questions that arise from examining models or a theory, to clarify and/or seek additional information and relationships.
- Ask questions to determine relationships, including quantitative relationships, between independent and dependent variables.
- Ask questions to clarify and refine a model, an explanation, or an engineering problem.
- Evaluate a question to determine if it is testable and relevant.
- Ask questions that can be investigated within the scope of the school laboratory, research facilities, or field (e.g., outdoor environment) with available resources and, when appropriate, frame a hypothesis based on a model or theory.
- Ask and/or evaluate questions that challenge the premise(s) of an argument, the interpretation of a data set, or the suitability of the design.
- Define a design problem that involves the development of a process or system with interacting components and criteria and constraints that may include social, technical, and/or environmental considerations.

Developing and Using Models for Grades 9–12

Modeling in 9–12 builds on K–8 experiences and progresses to using, synthesizing, and developing models to predict and show relationships among variables between systems and their components in the natural and designed world(s).

- Evaluate merits and limitations of two different models of the same proposed tool, process, mechanism, or system in order to select or revise a model that best fits the evidence or design criteria.
- Design a test of a model to ascertain its reliability.
- Develop, revise, and/or use a model based on evidence to illustrate and/or predict the relationships between systems or between components of a system.
- Develop and/or use multiple types of models to provide mechanistic accounts and/or predict phenomena, and move flexibly between model types based on merits and limitations.
- Develop a complex model that allows for manipulation and testing of a proposed process or system.
- Develop and/or use a model (including mathematical and computational) to generate data to support explanations, predict phenomena, analyze systems, and/or solve problems.

Planning and Carrying Out Investigations for Grades 9–12

Planning and carrying out investigations in 9–12 builds on K–8 experiences and progresses to include investigations that provide evidence for and test conceptual, mathematical, physical, and empirical models.

- Plan an investigation or test a design individually and collaboratively to produce data to serve as the basis for evidence as part of building and revising models, supporting explanations for phenomena, or testing solutions to problems. Consider possible variables or effects and evaluate the confounding investigation's design to ensure variables are controlled.
- Plan and conduct an investigation individually and collaboratively to produce data to serve as the basis for evidence, and in the design: decide on types, how much, and accuracy of data needed to produce reliable measurements and consider limitations on the precision of the data (e.g., number of trials, cost, risk, time), and refine the design accordingly.
- Plan and conduct an investigation or test a design solution in a safe and ethical manner including considerations of environmental, social, and personal impacts.
- Select appropriate tools to collect, record, analyze, and evaluate data.
- Make directional hypotheses that specify what happens to a dependent variable when an independent variable is manipulated.
- Manipulate variables and collect data about a complex model of a proposed process or system to identify failure points or improve performance relative to criteria for success or other variables.

Science and Engineering Practices (*continued*)

Analyzing and Interpreting Data for Grades 9–12

Analyzing data in 9–12 builds on K–8 experiences and progresses to introducing more detailed statistical analysis, the comparison of data sets for consistency, and the use of models to generate and analyze data.

- Analyze data using tools, technologies, and/or models (e.g., computational, mathematical) in order to make valid and reliable scientific claims or determine an optimal design solution.
- Apply concepts of statistics and probability (including determining function fits to data, slope, intercept, and correlation coefficient for linear fits) to scientific and engineering questions and problems, using digital tools when feasible.
- Consider limitations of data analysis (e.g., measurement error, sample selection) when analyzing and interpreting data.
- Compare and contrast various types of data sets (e.g., self-generated, archival) to examine consistency of measurements and observations.
- Evaluate the impact of new data on a working explanation and/or model of a proposed process or system.
- Analyze data to identify design features or characteristics of the components of a proposed process or system to optimize it relative to criteria for success.

Using Mathematics and Computational Thinking for Grades 9–12

Mathematical and computational thinking in 9–12 builds on K–8 experiences and progresses to using algebraic thinking and analysis, a range of linear and nonlinear functions including trigonometric functions, exponentials and logarithms, and computational tools for statistical analysis to analyze, represent, and model data. Simple computational simulations are created and used based on mathematical models of basic assumptions.

- Decide if qualitative or quantitative data are best to determine whether a proposed object or tool meets criteria for success.
- Create and/or revise a computational model or simulation of a phenomenon, designed device, process, or system.
- Use mathematical, computational, and/or algorithmic representations of phenomena or design solutions to describe and/or support claims and/or explanations.
- Apply techniques of algebra and functions to represent and solve scientific and engineering problems.
- Use simple limit cases to test mathematical expressions, computer programs, algorithms, or simulations of a process or system to see if a model "makes sense" by comparing the outcomes with what is known about the real world.
- Apply ratios, rates, percentages, and unit conversions in the context of complicated measurement problems involving quantities with derived or compound units (such as mg/mL, kg/m^3, acre-feet, etc.).

Constructing Explanations and Designing Solutions for Grades 9–12

Constructing explanations and designing solutions in 9–12 builds on K–8 experiences and progresses to explanations and designs that are supported by multiple and independent student-generated sources of evidence consistent with scientific ideas, principles, and theories.

- Make a quantitative and/or qualitative claim regarding the relationship between dependent and independent variables.
- Construct and revise an explanation based on valid and reliable evidence obtained from a variety of sources (including students' own investigations, models, theories, simulations, peer review) and the assumption that theories and laws that describe the natural world operate today as they did in the past and will continue to do so in the future.
- Apply scientific ideas, principles, and/or evidence to provide an explanation of phenomena and solve design problems, taking into account possible unanticipated effects.
- Apply scientific reasoning, theory, and/or models to link evidence to the claims to assess the extent to which the reasoning and data support the explanation or conclusion.
- Design, evaluate, and/or refine a solution to a complex real-world problem, based on scientific knowledge, student-generated sources of evidence, prioritized criteria, and trade-off considerations.

Science and Engineering Practices (*continued*)

Engaging in Argument From Evidence for Grades 9–12
Engaging in argument from evidence in 9–12 builds on K–8 experiences and progresses to using appropriate and sufficient evidence and scientific reasoning to defend and critique claims and explanations about the natural and designed world(s). Arguments may also come from current scientific or historical episodes in science. • Compare and evaluate competing arguments or design solutions in light of currently accepted explanations, new evidence, limitations (e.g., trade-offs), constraints, and ethical issues. • Evaluate the claims, evidence, and/or reasoning behind currently accepted explanations or solutions to determine the merits of arguments. • Respectfully provide and/or receive critiques on scientific arguments by probing reasoning and evidence and challenging ideas and conclusions, responding thoughtfully to diverse perspectives, and determining what additional information is required to resolve contradictions. • Construct, use, and/or present an oral and written argument or counterarguments based on data and evidence. • Make and defend a claim based on evidence about the natural world or the effectiveness of a design solution that reflects scientific knowledge and student-generated evidence. • Evaluate competing design solutions to a real-world problem based on scientific ideas and principles, empirical evidence, and/or logical arguments regarding relevant factors (e.g., economic, societal, environmental, ethical considerations).
Obtaining, Evaluating, and Communicating Information for Grades 9–12
Obtaining, evaluating, and communicating information in 9–12 builds on K–8 experiences and progresses to evaluating the validity and reliability of the claims, methods, and designs. • Critically read scientific literature adapted for classroom use to determine the central ideas or conclusions and/or to obtain scientific and/or technical information to summarize complex evidence, concepts, processes, or information presented in a text by paraphrasing them in simpler but still accurate terms. • Compare, integrate, and evaluate sources of information presented in different media or formats (e.g., visually, quantitatively) as well as in words in order to address a scientific question or solve a problem. • Gather, read, and evaluate scientific and/or technical information from multiple authoritative sources, assessing the evidence and usefulness of each source. • Evaluate the validity and reliability of and/or synthesize multiple claims, methods, and/or designs that appear in scientific and technical texts or media reports, verifying the data when possible. • Communicate scientific and/or technical information or ideas (e.g., about phenomena and/or the process of development and the design and performance of a proposed process or system) in multiple formats (including orally, graphically, textually, and mathematically).

Crosscutting Concepts and Connections to Engineering, Technology, and Applications of Science

Crosscutting Concepts for Grades 9–12	
Patterns	• Different patterns may be observed at each of the scales at which a system is studied and can provide evidence for causality in explanations of phenomena. • Classifications or explanations used at one scale may fail or need revision when information from smaller or larger scales is introduced; thus requiring improved investigations and experiments. • Patterns of performance of designed systems can be analyzed and interpreted to reengineer and improve the system. • Mathematical representations are needed to identify some patterns. • Empirical evidence is needed to identify patterns.
Cause and Effect: Mechanism and Prediction	• Empirical evidence is required to differentiate between cause and correlation and make claims about specific causes and effects. • Cause-and-effect relationships can be suggested and predicted for complex natural and human-designed systems by examining what is known about smaller scale mechanisms within the system. • Systems can be designed to cause a desired effect. • Changes in systems may have various causes that may not have equal effects.
Scale, Proportion, and Quantity	• The significance of a phenomenon is dependent on the scale, proportion, and quantity at which it occurs. • Some systems can only be studied indirectly as they are too small, too large, too fast, or too slow to observe directly. • Patterns observable at one scale may not be observable or exist at other scales. • Using the concept of orders of magnitude allows one to understand how a model at one scale relates to a model at another scale. • Algebraic thinking is used to examine scientific data and predict the effect of a change in one variable on another (e.g., linear growth vs. exponential growth).
Systems and System Models	• Systems can be designed to do specific tasks. • When investigating or describing a system, the boundaries and initial conditions of the system need to be defined and their inputs and outputs analyzed and described using models. • Models (e.g., physical, mathematical, computer models) can be used to simulate systems and interactions—including energy, matter, and information flows—within and between systems at different scales. • Models can be used to predict the behavior of a system, but these predictions have limited precision and reliability due to the assumptions and approximations inherent in models.
Energy and Matter: Flows, Cycles, and Conservation	• The total amount of energy and matter in closed systems is conserved. • Changes of energy and matter in a system can be described in terms of energy and matter flows into, out of, and within that system. • Energy cannot be created or destroyed—it only moves between one place and another place, between objects and/or fields, or between systems. • Energy drives the cycling of matter within and between systems. • In nuclear processes, atoms are not conserved, but the total number of protons plus neutrons is conserved.
Structure and Function	• Investigating or designing new systems or structures requires a detailed examination of the properties of different materials, the structures of different components, and connections of components to reveal its function and/or solve a problem. • The functions and properties of natural and designed objects and systems can be inferred from their overall structure, the way their components are shaped and used, and the molecular substructures of their various materials.

Crosscutting Concepts and Connections to Engineering, Technology, and Applications of Science (*continued*)

Crosscutting Concepts for Grades 9–12	
Stability and Change	• Much of science deals with constructing explanations of how things change and how they remain stable. • Change and rates of change can be quantified and modeled over very short or very long periods of time. Some system changes are irreversible. • Feedback (negative or positive) can stabilize or destabilize a system. • Systems can be designed for greater or lesser stability.
Connections to Engineering, Technology, and Applications of Science for Grades 9–12	
Interdependence of Science, Engineering, and Technology	• Science and engineering complement each other in the cycle known as research and development (R&D). • Many R&D projects may involve scientists, engineers, and others with wide ranges of expertise.
Influence of Science, Engineering, and Technology on Society and the Natural World	• Modern civilization depends on major technological systems, such as agriculture, health, water, energy, transportation, manufacturing, construction, and communications. • Engineers continuously modify these systems to increase benefits while decreasing costs and risks. • New technologies can have deep impacts on society and the environment, including some that were not anticipated. • Analysis of costs and benefits is a critical aspect of decisions about technology.

Connections to the Nature of Science

Understandings Most Closely Associated With Practices for Grades 9–12	
Scientific Investigations Use a Variety of Methods	• Science investigations use diverse methods and do not always use the same set of procedures to obtain data. • New technologies advance scientific knowledge. • Scientific inquiry is characterized by a common set of values that include logical thinking, precision, open-mindedness, objectivity, skepticism, replicability of results, and honest and ethical reporting of findings. • The discourse practices of science are organized around disciplinary domains that share exemplars for making decisions regarding the values, instruments, methods, models, and evidence to adopt and use. • Scientific investigations use a variety of methods, tools, and techniques to revise and produce new knowledge.
Scientific Knowledge Is Based on Empirical Evidence	• Science knowledge is based on empirical evidence. • Science disciplines share common rules of evidence used to evaluate explanations about natural systems. • Science includes the process of coordinating patterns of evidence with current theory. • Science arguments are strengthened by multiple lines of evidence supporting a single explanation.
Scientific Knowledge Is Open to Revision in Light of New Evidence	• Scientific explanations can be probabilistic. • Most scientific knowledge is quite durable but is, in principle, subject to change based on new evidence and/or reinterpretation of existing evidence. • Scientific argumentation is a mode of logical discourse used to clarify the strength of relationships between ideas and evidence that may result in revision of an explanation.
Science Models, Laws, Mechanisms, and Theories Explain Natural Phenomena	• Theories and laws provide explanations in science, but theories do not with time become laws or facts. • A scientific theory is a substantiated explanation of some aspect of the natural world, based on a body of facts that have been repeatedly confirmed through observation and experiment, and the science community validates each theory before it is accepted. If new evidence is discovered that the theory does not accommodate, the theory is generally modified in light of this new evidence. • Models, mechanisms, and explanations collectively serve as tools in the development of a scientific theory. • Laws are statements or descriptions of the relationships among observable phenomena. • Scientists often use hypotheses to develop and test theories and explanations.

Connections to the Nature of Science (*continued*)

Understandings Most Closely Associated With Crosscutting Concepts for Grades 9–12	
Science Is a Way of Knowing	• Science is both a body of knowledge that represents a current understanding of natural systems and the processes used to refine, elaborate, revise, and extend this knowledge. • Science is a unique way of knowing and there are other ways of knowing. • Science distinguishes itself from other ways of knowing through use of empirical standards, logical arguments, and skeptical review. • Science knowledge has a history that includes the refinement of, and changes to, theories, ideas, and beliefs over time.
Scientific Knowledge Assumes an Order and Consistency in Natural Systems	• Scientific knowledge is based on the assumption that natural laws operate today as they did in the past and they will continue to do so in the future. • Science assumes the universe is a vast single system in which basic laws are consistent.
Science Is a Human Endeavor	• Scientific knowledge is a result of human endeavor, imagination, and creativity. • Individuals and teams from many nations and cultures have contributed to science and to advances in engineering. • Technological advances have influenced the progress of science and science has influenced advances in technology. • Science and engineering are influenced by society, and society is influenced by science and engineering.
Science Addresses Questions About the Natural and Material World	• Not all questions can be answered by science. • Science and technology may raise ethical issues for which science, by itself, does not provide answers and solutions. • Science knowledge indicates what can happen in natural systems—not what should happen. The latter involves ethics, values, and human decisions about the use of knowledge. • Many decisions are not made using science alone, but rely on social and cultural contexts to resolve issues.

Performance Expectations and Disciplinary Core Ideas for Physical Science

Performance Expectations (PEs)	Disciplinary Core Ideas (DCIs)
HS-PS1-1. Use the periodic table as a model to predict the relative properties of elements based on the patterns of electrons in the outermost energy level of atoms. **Clarification Statement:** Examples of properties that could be predicted from patterns could include reactivity of metals, types of bonds formed, numbers of bonds formed, and reactions with oxygen. **Assessment Boundary:** Assessment is limited to main group elements. Assessment does not include quantitative understanding of ionization energy beyond relative trends.	**PS1.A. Structure and Properties of Matter** Each atom has a charged substructure consisting of a nucleus, which is made of protons and neutrons, surrounded by electrons. The periodic table orders elements horizontally by the number of protons in the atom's nucleus and places those with similar chemical properties in columns. The repeating patterns of this table reflect patterns of outer electron states. (HS-PS1-2) **PS2.B. Types of Interactions** Attraction and repulsion between electric charges at the atomic scale explain the structure, properties, and transformations of matter, as well as the contact forces between material objects. (HS-PS1-3), (HS-PS2-6)
HS-PS1-2. Construct and revise an explanation for the outcome of a simple chemical reaction based on the outermost electron states of atoms, trends in the periodic table, and knowledge of the patterns of chemical properties. **Clarification Statement:** Examples of chemical reactions could include the reaction of sodium and chlorine, of carbon and oxygen, or of carbon and hydrogen. **Assessment Boundary:** Assessment is limited to chemical reactions involving main group elements and combustion reactions.	**PS1.A. Structure and Properties of Matter** The periodic table orders elements horizontally by the number of protons in the atom's nucleus and places those with similar chemical properties in columns. The repeating patterns of this table reflect patterns of outer electron states. (HS-PS1-1) **PS1.B. Chemical Reactions** The fact that atoms are conserved, together with knowledge of the chemical properties of the elements involved, can be used to describe and predict chemical reactions. (HS-PS1-7)
HS-PS1-3. Plan and conduct an investigation to gather evidence to compare the structure of substances at the bulk scale to infer the strength of electrical forces between particles. **Clarification Statement:** Emphasis is on understanding the strengths of forces between particles, not on naming specific intermolecular forces (such as dipole-dipole). Examples of particles could include ions, atoms, molecules, and networked materials (such as graphite). Examples of bulk properties of substances could include the melting point and boiling point, vapor pressure, and surface tension. **Assessment Boundary:** Assessment does not include Raoult's law calculations of vapor pressure.	**PS2.B. Types of Interactions** Attraction and repulsion between electric charges at the atomic scale explain the structure, properties, and transformations of matter, as well as the contact forces between material objects. (HS-PS1-1), (HS-PS2-6) **PS1.A. Structure and Properties of Matter** The structure and interactions of matter at the bulk scale are determined by electrical forces within and between atoms.
HS-PS1-4. Develop a model to illustrate that the release or absorption of energy from a chemical reaction system depends upon the changes in total bond energy. **Clarification Statement:** Emphasis is on the idea that a chemical reaction is a system that affects the energy change. Examples of models could include molecular-level drawings and diagrams of reactions, graphs showing the relative energies of reactants and products, and representations showing energy is conserved. **Assessment Boundary:** Assessment does not include calculating the total bond energy changes during a chemical reaction from the bond energies of reactants and products.	**PS1.A. Structure and Properties of Matter** A stable molecule has less energy than the same set of atoms separated; one must provide at least this energy in order to take the molecule apart. **PS1.B. Chemical Reactions** Chemical processes, their rates, and whether or not energy is stored or released can be understood in terms of the collisions of molecules and the rearrangements of atoms into new molecules, with consequent changes in the sum of all bond energies in the set of molecules that are matched by changes in kinetic energy. (HS-PS1-5)

Performance Expectations and Disciplinary Core Ideas for Physical Science (*continued*)

Performance Expectations (PEs)	Disciplinary Core Ideas (DCIs)
HS-PS1-5. Apply scientific principles and evidence to provide an explanation about the effects of changing the temperature or concentration of the reacting particles on the rate at which a reaction occurs. **Clarification Statement:** Emphasis is on student reasoning that focuses on the number and energy of collisions between molecules. **Assessment Boundary:** Assessment is limited to simple reactions in which there are only two reactants; evidence from temperature, concentration, and rate data; and qualitative relationships between rate and temperature.	**PS1.B. Chemical Reactions** Chemical processes, their rates, and whether or not energy is stored or released can be understood in terms of the collisions of molecules and the rearrangements of atoms into new molecules, with consequent changes in the sum of all bond energies in the set of molecules that are matched by changes in kinetic energy. (HS-PS1-4)
HS-PS1-6. Refine the design of a chemical system by specifying a change in conditions that would produce increased amounts of products at equilibrium. **Clarification Statement:** Emphasis is on the application of Le Chatelier's principle and on refining designs of chemical reaction systems, including descriptions of the connection between changes made at the macroscopic level and what happens at the molecular level. Examples of designs could include different ways to increase product formation, including adding reactants or removing products. **Assessment Boundary:** Assessment is limited to specifying the change in only one variable at a time. Assessment does not include calculating equilibrium constants and concentrations.	**PS1.B. Chemical Reactions** In many situations, a dynamic and condition-dependent balance between a reaction and the reverse reaction determines the numbers of all types of molecules present. **ETS1.C. Optimizing the Design Solution** Criteria may need to be broken down into simpler ones that can be approached systematically, and decisions about the priority of certain criteria over others (trade-offs) may be needed. (HS-ETS1-2), (HS-PS2-3)
HS-PS1-7. Use mathematical representations to support the claim that atoms, and therefore mass, are conserved during a chemical reaction. **Clarification Statement:** Emphasis is on using mathematical ideas to communicate the proportional relationships between masses of atoms in the reactants and the products, and the translation of these relationships to the macroscopic scale using the mole as the conversion from the atomic to the macroscopic scale. Emphasis is on assessing students' use of mathematical thinking and not on memorization and rote application of problem-solving techniques **Assessment Boundary:** Assessment does not include complex chemical reactions.	**PS1.B. Chemical Reactions** The fact that atoms are conserved, together with knowledge of the chemical properties of the elements involved, can be used to describe and predict chemical reactions. (HS-PS1-2)
HS-PS1-8. Develop models to illustrate the changes in the composition of the nucleus of the atom and the energy released during the processes of fission, fusion, and radioactive decay. **Clarification Statement:** Emphasis is on simple qualitative models, such as pictures or diagrams, and on the scale of energy released in nuclear processes relative to other kinds of transformations. **Assessment Boundary:** Assessment does not include quantitative calculation of energy released. Assessment is limited to alpha, beta, and gamma radioactive decays.	**PS1.C. Nuclear Processes** Nuclear processes, including fusion, fission, and radioactive decays of unstable nuclei, involve release or absorption of energy. The total number of neutrons plus protons does not change in any nuclear process.

Performance Expectations and Disciplinary Core Ideas for Physical Science (*continued*)

Performance Expectations (PEs)	Disciplinary Core Ideas (DCIs)
HS-PS2-1. Analyze data to support the claim that Newton's second law of motion describes the mathematical relationship among the net force on a macroscopic object, its mass, and its acceleration. **Clarification Statement:** Examples of data could include tables or graphs of position or velocity as a function of time for objects subject to a net unbalanced force, such as a falling object, an object rolling down a ramp, or a moving object being pulled by a constant force. **Assessment Boundary:** Assessment is limited to one-dimensional motion and to macroscopic objects moving at non-relativistic speeds.	**PS2.A. Forces and Motion** Newton's second law accurately predicts changes in the motion of macroscopic objects.
HS-PS2-2. Use mathematical representations to support the claim that the total momentum of a system of objects is conserved when there is no net force on the system. **Clarification Statement:** Emphasis is on the quantitative conservation of momentum in interactions and the qualitative meaning of this principle. **Assessment Boundary:** Assessment is limited to systems of two macroscopic bodies moving in one dimension.	**PS2.A. Forces and Motion** Momentum is defined for a particular frame of reference; it is the mass times the velocity of the object. In any system, total momentum is always conserved. If a system interacts with objects outside itself, the total momentum of the system can change; however, any such change is balanced by changes in the momentum of objects outside the system. (HS-PS2-3)
HS-PS2-3. Apply scientific and engineering ideas to design, evaluate, and refine a device that minimizes the force on a macroscopic object during a collision. **Clarification Statement:** Examples of evaluation and refinement could include determining the success of the device at protecting an object from damage and modifying the design to improve it. Examples of a device could include a football helmet or a parachute. **Assessment Boundary:** Assessment is limited to qualitative evaluations and/or algebraic manipulations.	**PS2.A. Forces and Motion** If a system interacts with objects outside itself, the total momentum of the system can change; however, any such change is balanced by changes in the momentum of objects outside the system. (HS-PS2-2) **ETS1.A. Defining and Delimiting Engineering Problems** Criteria and constraints also include satisfying any requirements set by society, such as taking issues of risk mitigation into account, and they should be quantified to the extent possible and stated in such a way that one can tell if a given design meets them. (HS-ETS1-1) **ETS1.C. Optimizing the Design Solution** Criteria may need to be broken down into simpler ones that can be approached systematically, and decisions about the priority of certain criteria over others (trade-offs) may be needed. (HS-ETS1-2), (HS-PS1-6)
HS-PS2-4. Use mathematical representations of Newton's law of gravitation and Coulomb's law to describe and predict the gravitational and electrostatic forces between objects. **Clarification Statement:** Emphasis is on both quantitative and conceptual descriptions of gravitational and electric fields. **Assessment Boundary:** Assessment is limited to systems with two objects.	**PS2.B. Types of Interactions** Newton's law of universal gravitation and Coulomb's law provide the mathematical models to describe and predict the effects of gravitational and electrostatic forces between distant objects. Forces at a distance are explained by fields (gravitational, electric, and magnetic) permeating space that can transfer energy through space. Magnets or electric currents cause magnetic fields; electric charges or changing magnetic fields cause electric fields. (HS-PS2-5)

Performance Expectations and Disciplinary Core Ideas for Physical Science (*continued*)

Performance Expectations (PEs)	Disciplinary Core Ideas (DCIs)
HS-PS2-5. Plan and conduct an investigation to provide evidence that an electric current can produce a magnetic field and that a changing magnetic field can produce an electric current. **Assessment Boundary:** Assessment is limited to designing and conducting investigations with provided materials and tools.	**PS2.B. Types of Interactions** Forces at a distance are explained by fields (gravitational, electric, and magnetic) permeating space that can transfer energy through space. Magnets or electric currents cause magnetic fields; electric charges or changing magnetic fields cause electric fields. (HS-PS2-4) **PS3.A. Definitions of Energy** "Electrical energy" may mean energy stored in a battery or energy transmitted by electric currents.
HS-PS2-6. Communicate scientific and technical information about why the molecular-level structure is important in the functioning of designed materials. **Clarification Statement:** Emphasis is on the attractive and repulsive forces that determine the functioning of the material. Examples could include why electrically conductive materials are often made of metal, flexible but durable materials are made up of long chained molecules, and pharmaceuticals are designed to interact with specific receptors. **Assessment Boundary:** Assessment is limited to provided molecular structures of specific designed materials.	**PS2.B. Types of Interactions** Attraction and repulsion between electric charges at the atomic scale explain the structure, properties, and transformations of matter, as well as the contact forces between material objects. (HS-PS1-1), (HS-PS1-3) **PS1.A. Structure and Properties of Matter** The structure and interactions of matter at the bulk scale are determined by electrical forces within and between atoms. (HS-PS1-3)
HS-PS3-1. Create a computational model to calculate the change in the energy of one component in a system when the change in energy of the other component(s) and energy flows in and out of the system are known. **Clarification Statement:** Emphasis is on explaining the meaning of mathematical expressions used in the model. **Assessment Boundary:** Assessment is limited to basic algebraic expressions or computations; to systems of two or three components; and to thermal energy, kinetic energy, and/or the energies in gravitational, magnetic, or electric fields.	**PS3.A. Definitions of Energy** Energy is a quantitative property of a system that depends on the motion and interactions of matter and radiation within that system. That there is a single quantity called energy is due to the fact that a system's total energy is conserved, even as, within the system, energy is continually transferred from one object to another and between its various possible forms. (HS-PS3-2) **PS3.B. Conservation of Energy and Energy Transfer** Conservation of energy means that the total change of energy in any system is always equal to the total energy transferred into or out of the system. Energy cannot be created or destroyed, but it can be transported from one place to another and transferred between systems. (HS-PS3-4) Mathematical expressions, which quantify how the stored energy in a system depends on its configuration (e.g., relative positions of charged particles, compression of a spring) and how kinetic energy depends on mass and speed, allow the concept of conservation of energy to be used to predict and describe system behavior. The availability of energy limits what can occur in any system.

Performance Expectations and Disciplinary Core Ideas for Physical Science (*continued*)

Performance Expectations (PEs)	Disciplinary Core Ideas (DCIs)
HS-PS3-2. Develop and use models to illustrate that energy at the macroscopic scale can be accounted for as a combination of energy associated with the motions of particles (objects) and energy associated with the relative positions of particles (objects) **Clarification Statement:** Examples of phenomena at the macroscopic scale could include the conversion of kinetic energy to thermal energy, the energy stored due to position of an object above the Earth, and the energy stored between two electrically charged plates. Examples of models could include diagrams, drawings, descriptions, and computer simulations.	**PS3.A. Definitions of Energy** Energy is a quantitative property of a system that depends on the motion and interactions of matter and radiation within that system. That there is a single quantity called energy is due to the fact that a system's total energy is conserved, even as, within the system, energy is continually transferred from one object to another and between its various possible forms. (HS-PS3-1) At the macroscopic scale, energy manifests itself in multiple ways, such as in motion, sound, light, and thermal energy. These relationships are better understood at the microscopic scale, at which all of the different manifestations of energy can be modeled as a combination of energy associated with the motion of particles and energy associated with the configuration (relative position of the particles). In some cases the relative position energy can be thought of as stored in fields (which mediate interactions between particles). This last concept includes radiation, a phenomenon in which energy stored in fields moves across space.
HS-PS3-3. Design, build, and refine a device that works within given constraints to convert one form of energy into another form of energy. **Clarification Statement:** Emphasis is on both qualitative and quantitative evaluations of devices. Examples of devices could include Rube Goldberg devices, wind turbines, solar cells, solar ovens, and generators. Examples of constraints could include use of renewable energy forms and efficiency. **Assessment Boundary:** Assessment for quantitative evaluations is limited to total output for a given input. Assessment is limited to devices constructed with materials provided to students.	**PS3.A. Definitions of Energy** At the macroscopic scale, energy manifests itself in multiple ways, such as in motion, sound, light, and thermal energy. **PS3.D. Energy in Chemical Processes and Everyday Life** Although energy cannot be destroyed, it can be converted to less useful forms—for example, to thermal energy in the surrounding environment. (HS-PS3-4) **ETS1.A. Defining and Delimiting Engineering Problems** Criteria and constraints also include satisfying any requirements set by society, such as taking issues of risk mitigation into account, and they should be quantified to the extent possible and stated in such a way that one can tell if a given design meets them. (HS-ETS1-1)
HS-PS3-4. Plan and conduct an investigation to provide evidence that the transfer of thermal energy when two components of different temperature are combined within a closed system results in a more uniform energy distribution among the components in the system (second law of thermodynamics). **Clarification Statement:** Emphasis is on analyzing data from student investigations and using mathematical thinking to describe the energy changes both quantitatively and conceptually. Examples of investigations could include mixing liquids at different initial temperatures or adding objects at different temperatures to water. **Assessment Boundary:** Assessment is limited to investigations based on materials and tools provided to students.	**PS3.D. Energy in Chemical Processes and Everyday Life** Although energy cannot be destroyed, it can be converted to less useful forms—for example, to thermal energy in the surrounding environment. (HS-PS3-3) **PS3.B. Conservation of Energy and Energy Transfer** Energy cannot be created or destroyed, but it can be transported from one place to another and transferred between systems. (HS-PS3-1) Uncontrolled systems always evolve toward more stable states—that is, toward more uniform energy distribution (e.g., water flows downhill, objects hotter than their surrounding environment cool down).

Performance Expectations and Disciplinary Core Ideas for Physical Science (*continued*)

Performance Expectations (PEs)	Disciplinary Core Ideas (DCIs)
HS-PS3-5. Develop and use a model of two objects interacting through electric or magnetic fields to illustrate the forces between objects and the changes in energy of the objects due to the interaction. **Clarification Statement:** Examples of models could include drawings, diagrams, and texts, such as drawings of what happens when two charges of opposite polarity are near each other. **Assessment Boundary:** Assessment is limited to systems containing two objects.	**PS3.C. Relationship Between Energy and Forces** When two objects interacting through a field change relative position, the energy stored in the field is changed.
HS-PS4-1. Use mathematical representations to support a claim regarding relationships among the frequency, wavelength, and speed of waves traveling in various media. **Clarification Statement:** Examples of data could include electromagnetic radiation traveling in a vacuum and glass, sound waves traveling through air and water, and seismic waves traveling through the Earth. **Assessment Boundary:** Assessment is limited to algebraic relationships and describing those relationships qualitatively.	**PS4.A. Wave Properties** The wavelength and frequency of a wave are related to one another by the speed of travel of the wave, which depends on the type of wave and the medium through which it is passing.
HS-PS4-2. Evaluate questions about the advantages of using a digital transmission and storage of information. **Clarification Statement:** Examples of advantages could include that digital information is stable because it can be stored reliably in computer memory, transferred easily, and copied and shared rapidly. Disadvantages could include issues of easy deletion, security, and theft.	**PS4.A. Wave Properties** Information can be digitized (e.g., a picture stored as the values of an array of pixels); in this form, it can be stored reliably in computer memory and sent over long distances as a series of wave pulses. (HS-PS4-5)
HS-PS4-3. Evaluate the claims, evidence, and reasoning behind the idea that electromagnetic radiation can be described either by a wave model or a particle model, and that for some situations one model is more useful than the other. **Clarification Statement:** Emphasis is on how the experimental evidence supports the claim and how a theory is generally modified in light of new evidence. Examples of a phenomenon could include resonance, interference, diffraction, and photoelectric effect. **Assessment Boundary:** Assessment does not include using quantum theory.	**PS4.A. Wave Properties** [From the 3–5 grade band endpoints] Waves can add or cancel one another as they cross, depending on their relative phase (i.e., relative position of peaks and troughs of the waves), but they emerge unaffected by each other. (Boundary: The discussion at this grade level is qualitative only; it can be based on the fact that two different sounds can pass a location in different directions without getting mixed up.) **PS4.B. Electromagnetic Radiation** Electromagnetic radiation (e.g., radio, microwaves, light) can be modeled as a wave of changing electric and magnetic fields or as particles called photons. The wave model is useful for explaining many features of electromagnetic radiation, and the particle model explains other features.

Performance Expectations and Disciplinary Core Ideas for Physical Science (*continued*)

Performance Expectations (PEs)	Disciplinary Core Ideas (DCIs)
HS-PS4-4. Evaluate the validity and reliability of claims in published materials of the effects that different frequencies of electromagnetic radiation have when absorbed by matter. **Clarification Statement:** Emphasis is on the idea that photons associated with different frequencies of light have different energies, and the damage to living tissue from electromagnetic radiation depends on the energy of the radiation. Examples of published materials could include trade books, magazines, web resources, videos, and other passages that may reflect bias. **Assessment Boundary:** Assessment is limited to qualitative descriptions.	**PS4.B. Electromagnetic Radiation** When light or longer wavelength electromagnetic radiation is absorbed in matter, it is generally converted into thermal energy (heat). Shorter wavelength electromagnetic radiation (ultraviolet, x-rays, gamma rays) can ionize atoms and cause damage to living cells.
HS-PS4-5. Communicate technical information about how some technological devices use the principles of wave behavior and wave interactions with matter to transmit and capture information and energy. **Clarification Statement:** Examples could include solar cells capturing light and converting it to electricity; medical imaging; and communications technology. **Assessment Boundary:** Assessments are limited to qualitative information. Assessments do not include band theory.	**PS3.D. Energy in Chemical Processes and Everyday Life** Solar cells are human-made devices that likewise capture the Sun's energy and produce electrical energy. **PS4.A. Wave Properties** Information can be digitized (e.g., a picture stored as the values of an array of pixels); in this form, it can be stored reliably in computer memory and sent over long distances as a series of wave pulses. (HS-PS4-2) **PS4.B. Electromagnetic Radiation** Photoelectric materials emit electrons when they absorb light of a high-enough frequency. **PS4.C. Information Technologies and Instrumentation** Multiple technologies based on the understanding of waves and their interactions with matter are part of everyday experiences in the modern world (e.g., medical imaging, communications, scanners) and in scientific research. They are essential tools for producing, transmitting, and capturing signals and for storing and interpreting the information contained in them.

Performance Expectations and Disciplinary Core Ideas for Life Science

Performance Expectations (PEs)	Disciplinary Core Ideas (DCIs)
HS-LS1-1. Construct an explanation based on evidence for how the structure of DNA determines the structure of proteins which carry out the essential functions of life through systems of specialized cells. **Assessment Boundary:** Assessment does not include identification of specific cell or tissue types, whole body systems, specific protein structures and functions, or the biochemistry of protein synthesis.	**LS1.A. Structure and Function** Systems of specialized cells within organisms help them perform the essential functions of life. All cells contain genetic information in the form of DNA molecules. Genes are regions in the DNA that contain the instructions that code for the formation of proteins. (HS-LS3-1)
HS-LS1-2. Develop and use a model to illustrate the hierarchical organization of interacting systems that provide specific functions within multicellular organisms. **Clarification Statement:** Emphasis is on functions at the organism system level such as nutrient uptake, water delivery, and organism movement in response to neural stimuli. An example of an interacting system could be an artery depending on the proper function of elastic tissue and smooth muscle to regulate and deliver the proper amount of blood within the circulatory system. **Assessment Boundary:** Assessment does not include interactions and functions at the molecular or chemical reaction level.	**LS1.A. Structure and Function** Multicellular organisms have a hierarchical structural organization, in which any one system is made up of numerous parts and is itself a component of the next level. (HS-PS4-5)
HS-LS1-3. Plan and conduct an investigation to provide evidence that feedback mechanisms maintain homeostasis. **Clarification Statement:** Examples of investigations could include heart rate response to exercise, stomate response to moisture and temperature, and root development in response to water levels. **Assessment Boundary:** Assessment does not include the cellular processes involved in the feedback mechanism.	**LS1.A. Structure and Function** Feedback mechanisms maintain a living system's internal conditions within certain limits and mediate behaviors, allowing it to remain alive and functional even as external conditions change within some range. Feedback mechanisms can encourage (through positive feedback) or discourage (negative feedback) what is going on inside the living system.
HS-LS1-4. Use a model to illustrate the role of cellular division (mitosis) and differentiation in producing and maintaining complex organisms. **Assessment Boundary:** Assessment does not include specific gene control mechanisms or rote memorization of the steps of mitosis.	**LS1.B. Growth and Development of Organisms** In multicellular organisms individual cells grow and then divide via a process called mitosis, thereby allowing the organism to grow. The organism begins as a single cell (fertilized egg) that divides successively to produce many cells, with each parent cell passing identical genetic material (two variants of each chromosome pair) to both daughter cells. Cellular division and differentiation produce and maintain a complex organism, composed of systems of tissues and organs that work together to meet the needs of the whole organism.
HS-LS1-5. Use a model to illustrate how photosynthesis transforms light energy into stored chemical energy. **Clarification Statement:** Emphasis is on illustrating inputs and outputs of matter and the transfer and transformation of energy in photosynthesis by plants and other photosynthesizing organisms. Examples of models could include diagrams, chemical equations, and conceptual models. **Assessment Boundary:** Assessment does not include specific biochemical steps.	**LS1.C. Organization for Matter and Energy Flow in Organisms** The process of photosynthesis converts light energy to stored chemical energy by converting carbon dioxide plus water into sugars plus released oxygen.

Performance Expectations and Disciplinary Core Ideas for Life Science (*continued*)

Performance Expectations (PEs)	Disciplinary Core Ideas (DCIs)
HS-LS1-6. Construct and revise an explanation based on evidence for how carbon, hydrogen, and oxygen from sugar molecules may combine with other elements to form amino acids and/or other large carbon-based molecules. **Clarification Statement:** Emphasis is on using evidence from models and simulations to support explanations. **Assessment Boundary:** Assessment does not include the details of the specific chemical reactions or identification of macromolecules.	**LS1.C. Organization for Matter and Energy Flow in Organisms** The sugar molecules thus formed contain carbon, hydrogen, and oxygen: their hydrocarbon backbones are used to make amino acids and other carbon-based molecules that can be assembled into larger molecules (such as proteins or DNA), used for example to form new cells. As matter and energy flow through different organizational levels of living systems, chemical elements are recombined in different ways to form different products.
HS-LS1-7. Use a model to illustrate that cellular respiration is a chemical process whereby the bonds of food molecules and oxygen molecules are broken and the bonds in new compounds are formed, resulting in a net transfer of energy. **Clarification Statement:** Emphasis is on the conceptual understanding of the inputs and outputs of the process of cellular respiration. **Assessment Boundary:** Assessment should not include identification of the steps or specific processes involved in cellular respiration.	**LS1.C. Organization for Matter and Energy Flow in Organisms** As matter and energy flow through different organizational levels of living systems, chemical elements are recombined in different ways to form different products. **LS1.C. Organization for Matter and Energy Flow in Organisms** As a result of these chemical reactions, energy is transferred from one system of interacting molecules to another. Cellular respiration is a chemical process in which the bonds of food molecules and oxygen molecules are broken and new compounds are formed that can transport energy to muscles. Cellular respiration also releases the energy needed to maintain body temperature despite ongoing energy transfer to the surrounding environment.
HS-LS2-1. Use mathematical and/or computational representations to support explanations of factors that affect carrying capacity of ecosystems at different scales. **Clarification Statement:** Emphasis is on quantitative analysis and comparison of the relationships among interdependent factors including boundaries, resources, climate, and competition. Examples of mathematical comparisons could include graphs, charts, histograms, and population changes gathered from simulations or historical data sets. **Assessment Boundary:** Assessment does not include deriving mathematical equations to make comparisons.	**LS2.A. Interdependent Relationships in Ecosystems** Ecosystems have carrying capacities, which are limits to the numbers of organisms and populations they can support. These limits result from factors as the availability of living and nonliving resources and from challenges such as predation, competition, and disease. Organisms would have the capacity to produce populations of great size were it not for the fact that environments and resources are finite. This fundamental tension affects the abundance (number of individuals) of species in any given ecosystem. (HS-LS2-2)

Performance Expectations and Disciplinary Core Ideas for Life Science (*continued*)

Performance Expectations (PEs)	Disciplinary Core Ideas (DCIs)
HS-LS2-2. Use mathematical representations to support and revise explanations based on evidence about factors affecting biodiversity and populations in ecosystems of different scales. **Clarification Statement:** Examples of mathematical representations include finding the average, determining trends, and using graphical comparisons of multiple sets of data. **Assessment Boundary:** Assessment is limited to provided data.	**LS2.A. Interdependent Relationships in Ecosystems** Ecosystems have carrying capacities, which are limits to the numbers of organisms and populations they can support. These limits result from such factors as the availability of living and nonliving resources and from challenges such as predation, competition, and disease. Organisms would have the capacity to produce populations of great size were it not for the fact that environments and resources are finite. This fundamental tension affects the abundance (number of individuals) of species in any given ecosystem. (HS-LS2-1) **LS2.C. Ecosystem Dynamics, Functioning, and Resilience** A complex set of interactions within an ecosystem can keep its numbers and types of organisms relatively constant over long periods of time under stable conditions. If a modest biological or physical disturbance to an ecosystem occurs, it may return to its more or less original status (i.e., the ecosystem is resilient), as opposed to becoming a very different ecosystem. Extreme fluctuations in conditions or the size of any population, however, can challenge the functioning of ecosystems in terms of resources and habitat availability. (HS-LS2-6)
HS-LS2-3. Construct and revise an explanation based on evidence for the cycling of matter and flow of energy in aerobic and anaerobic conditions. **Clarification Statement:** Emphasis is on conceptual understanding of the role of aerobic and anaerobic respiration in different environments. **Assessment Boundary:** Assessment does not include the specific chemical processes of either aerobic or anaerobic respiration.	**LS2.B. Cycles of Matter and Energy Transfer in Ecosystems** Photosynthesis and cellular respiration (including anaerobic processes) provide most of the energy for life processes. (HS-LS2-5)
HS-LS2-4. Use mathematical representations to support claims for the cycling of matter and flow of energy among organisms in an ecosystem **Clarification Statement:** Emphasis is on using a mathematical model of stored energy in biomass to describe the transfer of energy from one trophic level to another and that matter and energy are conserved as matter cycles and energy flows through ecosystems. Emphasis is on atoms and molecules such as carbon, oxygen, hydrogen, and nitrogen being conserved as they move through an ecosystem. **Assessment Boundary:** Assessment is limited to proportional reasoning to describe the cycling of matter and flow of energy.	**LS2.B. Cycles of Matter and Energy Transfer in Ecosystems** Plants or algae form the lowest level of the food web. At each link upward in a food web, only a small fraction of the matter consumed at the lower level is transferred upward, to produce growth and release energy in cellular respiration at the higher level. Given this inefficiency, there are generally fewer organisms at higher levels of a food web. Some matter reacts to release energy for life functions, some matter is stored in newly made structures, and much is discarded. The chemical elements that make up the molecules of organisms pass through food webs and into and out of the atmosphere and soil, and they are combined and recombined in different ways. At each link in an ecosystem, matter and energy are conserved.

Performance Expectations and Disciplinary Core Ideas for Life Science (*continued*)

Performance Expectations (PEs)	Disciplinary Core Ideas (DCIs)
HS-LS2-5. Develop a model to illustrate the role of photosynthesis and cellular respiration in the cycling of carbon among the biosphere, atmosphere, hydrosphere, and geosphere. **Clarification Statement:** Examples of models could include simulations and mathematical models. **Assessment Boundary:** Assessment does not include the specific chemical steps of photosynthesis and respiration.	**LS2.B. Cycles of Matter and Energy Transfer in Ecosystems** Photosynthesis and cellular respiration are important components of the carbon cycle, in which carbon is exchanged among the biosphere, atmosphere, oceans, and geosphere through chemical, physical, geologic, and biological processes. **PS3.D. Energy in Chemical Processes and Everyday Life** The main way that solar energy is captured and stored on Earth is through the complex chemical process known as photosynthesis.
HS-LS2-6. Evaluate the claims, evidence, and reasoning that the complex interactions in ecosystems maintain relatively consistent numbers and types of organisms in stable conditions, but changing conditions may result in a new ecosystem. **Clarification Statement:** Examples of changes in ecosystem conditions could include modest biological or physical changes, such as moderate hunting or a seasonal flood, and extreme changes, such as volcanic eruption or sea level rise.	**LS2.C. Ecosystem Dynamics, Functioning, and Resilience** A complex set of interactions within an ecosystem can keep its numbers and types of organisms relatively constant over long periods of time under stable conditions. If a modest biological or physical disturbance to an ecosystem occurs, it may return to its more or less original status (i.e., the ecosystem is resilient), as opposed to becoming a very different ecosystem. Extreme fluctuations in conditions or the size of any population, however, can challenge the functioning of ecosystems in terms of resources and habitat availability. (HS-LS2-2)
HS-LS2-7. Design, evaluate, and refine a solution for reducing the impacts of human activities on the environment and biodiversity. **Clarification Statement:** Examples of human activities can include urbanization, building dams, and dissemination of invasive species.	**LS2.C. Ecosystem Dynamics, Functioning, and Resilience** Anthropogenic changes (induced by human activity) in the environment—including habitat destruction, pollution, introduction of invasive species, overexploitation, and climate change—can disrupt an ecosystem and threaten the survival of some species. **LS4.D. Biodiversity and Humans** Biodiversity is increased by the formation of new species (speciation) and decreased by the loss of species (extinction). Humans depend on the living world for the resources and other benefits provided by biodiversity. But human activity is also having adverse impacts on biodiversity through overpopulation, overexploitation, habitat destruction, pollution, introduction of invasive species, and climate change. Thus, sustaining biodiversity so that ecosystem functioning and productivity are maintained is essential to supporting and enhancing life on Earth. Sustaining biodiversity also aids humanity by preserving landscapes of recreational or inspirational value. (HS-LS4-6) **ETS1.B. Developing Possible Solutions** When evaluating solutions it is important to take into account a range of constraints including cost, safety, reliability, and aesthetics and to consider social, cultural, and environmental impacts.

Performance Expectations and Disciplinary Core Ideas for Life Science (*continued*)

Performance Expectations (PEs)	Disciplinary Core Ideas (DCIs)
HS-LS2-8. Evaluate the evidence for the role of group behavior on individuals' and species' chances to survive and reproduce. **Clarification Statement:** Emphasis is on (1) distinguishing between group and individual behavior, (2) identifying evidence supporting the outcomes of group behavior, and (3) developing logical and reasonable arguments based on evidence. Examples of group behaviors could include flocking, schooling, herding, and cooperative behaviors such as hunting, migrating, and swarming.	**LS2.D. Social Interactions and Group Behavior** Group behavior has evolved because membership can increase the chances of survival for individuals and their genetic relatives.
HS-LS3-1. Ask questions to clarify relationships about the role of DNA and chromosomes in coding the instructions for characteristic traits passed from parents to offspring. **Assessment Boundary:** Assessment does not include the phases of meiosis or the biochemical mechanism of specific steps in the process.	**LS3.A. Inheritance of Traits** Each chromosome consists of a single very long DNA molecule, and each gene on the chromosome is a particular segment of that DNA. The instructions for forming species' characteristics are carried in DNA. All cells in an organism have the same genetic content, but the genes used (expressed) by the cell may be regulated in different ways. Not all DNA codes for a protein; some segments of DNA are involved in regulatory or structural functions, and some have no as-yet known function. **LS1.A. Structure and Function** All cells contain genetic information in the form of DNA molecules. Genes are regions in the DNA that contain the instructions that code for the formation of proteins. (HS-LS1-1)
HS-LS3-2. Make and defend a claim based on evidence that inheritable genetic variations may result from (1) new genetic combinations through meiosis, (2) viable errors occurring during replication, and/or (3) mutations caused by environmental factors. **Clarification Statement:** Emphasis is on using data to support arguments for the way variation occurs. **Assessment Boundary:** Assessment does not include the phases of meiosis or the biochemical mechanism of specific steps in the process.	**LS3.B. Variation of Traits** In sexual reproduction, chromosomes can sometimes swap sections during the process of meiosis (cell division), thereby creating new genetic combinations and thus more genetic variation. Although DNA replication is tightly regulated and remarkably accurate, errors do occur and result in mutations, which are also a source of genetic variation. Environmental factors can also cause mutations in genes, and viable mutations are inherited. Environmental factors also affect expression of traits, and hence affect the probability of occurrences of traits in a population. Thus, the variation and distribution of traits observed depends on both genetic and environmental factors. (HS-LS3-3)
HS-LS3-3. Apply concepts of statistics and probability to explain the variation and distribution of expressed traits in a population. **Clarification Statement:** Emphasis is on the use of mathematics to describe the probability of traits as it relates to genetic and environmental factors in the expression of traits. **Assessment Boundary:** Assessment does not include Hardy-Weinberg calculations.	**LS3.B. Variation of Traits** Environmental factors also affect expression of traits, and hence affect the probability of occurrences of traits in a population. Thus, the variation and distribution of traits observed depends on both genetic and environmental factors. (HS-LS3-2)

Performance Expectations and Disciplinary Core Ideas for Life Science (*continued*)

Performance Expectations (PEs)	Disciplinary Core Ideas (DCIs)
HS-LS4-1. Communicate scientific information that common ancestry and biological evolution are supported by multiple lines of empirical evidence. **Clarification Statement:** Emphasis is on a conceptual understanding of the role each line of evidence has relating to common ancestry and biological evolution. Examples of evidence could include similarities in DNA sequences, anatomical structures, and order of appearance of structures in embryological development.	**LS4.A. Evidence of Common Ancestry and Diversity** Genetic information provides evidence of evolution. DNA sequences vary among species, but there are many overlaps; in fact, the ongoing branching that produces multiple lines of descent can be inferred by comparing the DNA sequences of different organisms. Such information is also derivable from the similarities and differences in amino acid sequences and from anatomical and embryological evidence.
HS-LS4-2. Construct an explanation based on evidence that the process of evolution primarily results from four factors: (1) the potential for a species to increase in number, (2) the heritable genetic variation of individuals in a species due to mutation and sexual reproduction, (3) competition for limited resources, and (4) the proliferation of those organisms that are better able to survive and reproduce in the environment. **Clarification Statement:** Emphasis is on using evidence to explain the influence each of the four factors has on number of organisms, behaviors, morphology, or physiology in terms of ability to compete for limited resources and subsequent survival of individuals and adaptation of species. Examples of evidence could include mathematical models such as simple distribution graphs and proportional reasoning. **Assessment Boundary:** Assessment does not include other mechanisms of evolution, such as genetic drift, gene flow through migration, and co-evolution.	**LS4.B. Natural Selection** Natural selection occurs only if there is both (1) variation in the genetic information between organisms in a population and (2) variation in the expression of that genetic information—that is, trait variation—that leads to differences in performance among individuals. (HS-LS4-3) **LS4.C. Adaptation** Evolution is a consequence of the interaction of four factors: (1) the potential for a species to increase in number, (2) the genetic variation of individuals in a species due to mutation and sexual reproduction, (3) competition for an environment's limited supply of the resources that individuals need in order to survive and reproduce, and (4) the ensuing proliferation of those organisms that are better able to survive and reproduce in that environment.
HS-LS4-3. Apply concepts of statistics and probability to support explanations that organisms with an advantageous heritable trait tend to increase in proportion to organisms lacking this trait. **Clarification Statement:** Emphasis is on analyzing shifts in numerical distribution of traits and using these shifts as evidence to support explanations. **Assessment Boundary:** Assessment is limited to basic statistical and graphical analysis. Assessment does not include allele frequency calculations.	**LS4.B. Natural Selection** Natural selection occurs only if there is both (1) variation in the genetic information between organisms in a population and (2) variation in the expression of that genetic information—that is, trait variation—that leads to differences in performance among individuals. (HS-LS4-2) The traits that positively affect survival are more likely to be reproduced, and thus are more common in the population. **LS4.C. Adaptation** Natural selection leads to adaptation, that is, to a population dominated by organisms that are anatomically, behaviorally, and physiologically well suited to survive and reproduce in a specific environment. That is, the differential survival and reproduction of organisms in a population that have an advantageous heritable trait lead to an increase in the proportion of individuals in future generations that have the trait and to a decrease in the proportion of individuals that do not. (HS-LS4-4) **LS4.C. Adaptation** Adaptation also means that the distribution of traits in a population can change when conditions change.

Performance Expectations and Disciplinary Core Ideas for Life Science (*continued*)

Performance Expectations (PEs)	Disciplinary Core Ideas (DCIs)
HS-LS4-4. Construct an explanation based on evidence for how natural selection leads to adaptation of populations. **Clarification Statement:** Emphasis is on using data to provide evidence for how specific biotic and abiotic differences in ecosystems (such as ranges of seasonal temperature, long-term climate change, acidity, light, geographic barriers, or evolution of other organisms) contribute to a change in gene frequency over time, leading to adaptation of populations.	**LS4.C. Adaptation** Natural selection leads to adaptation, that is, to a population dominated by organisms that are anatomically, behaviorally, and physiologically well suited to survive and reproduce in a specific environment. That is, the differential survival and reproduction of organisms in a population that have an advantageous heritable trait lead to an increase in the proportion of individuals in future generations that have the trait and to a decrease in the proportion of individuals that do not. (HS-LS4-3)
HS-LS4-5. Evaluate the evidence supporting claims that changes in environmental conditions may result in (1) increases in the number of individuals of some species, (2) the emergence of new species over time, and (3) the extinction of other species. **Clarification Statement:** Emphasis is on determining cause-and-effect relationships for how changes to the environment such as deforestation, fishing, application of fertilizers, drought, flood, and the rate of change of the environment affect distribution or disappearance of traits in species.	**LS4.C. Adaptation** Changes in the physical environment, whether naturally occurring or human induced, have thus contributed to the expansion of some species, the emergence of new distinct species as populations diverge under different conditions, and the decline–and sometimes the extinction–of some species. **LS4.C. Adaptation** Species become extinct because they can no longer survive and reproduce in their altered environment. If members cannot adjust to change that is too fast or drastic, the opportunity for the species' evolution is lost.
HS-LS4-6. Create or revise a simulation to test a solution to mitigate adverse impacts of human activity on biodiversity. **Clarification Statement:** Emphasis is on designing solutions for a proposed problem related to threatened or endangered species, or to genetic variation of organisms for multiple species.	**LS4.C. Adaptation** Changes in the physical environment, whether naturally occurring or human induced, have thus contributed to the expansion of some species, the emergence of new distinct species as populations diverge under different conditions, and the decline–and sometimes the extinction–of some species. (HS-LS4-5) **LS4.D. Biodiversity and Humans** Humans depend on the living world for the resources and other benefits provided by biodiversity. But human activity is also having adverse impacts on biodiversity through overpopulation, overexploitation, habitat destruction, pollution, introduction of invasive species, and climate change. Thus, sustaining biodiversity so that ecosystem functioning and productivity are maintained is essential to supporting and enhancing life on Earth. Sustaining biodiversity also aids humanity by preserving landscapes of recreational or inspirational value. (HS-LS2-7) **ETS1.B. Developing Possible Solutions** When evaluating solutions it is important to take into account a range of constraints including cost, safety, reliability, and aesthetics and to consider social, cultural, and environmental impacts. (HS-ETS1-3) Both physical models and computers can be used in various ways to aid in the engineering design process. Computers are useful for a variety of purposes, such as running simulations to test different ways of solving a problem or to see which one is most efficient or economical; and in making a persuasive presentation to a client about how a given design will meet his or her needs. (HS-ETS1-4)

Performance Expectations and Disciplinary Core Ideas for Earth and Space Science

Performance Expectations (PEs)	Disciplinary Core Ideas (DCIs)
HS-ESS1-1. Develop a model based on evidence to illustrate the life span of the Sun and the role of nuclear fusion in the Sun's core to release energy in the form of radiation. **Clarification Statement:** Emphasis is on the energy transfer mechanisms that allow energy from nuclear fusion in the Sun's core to reach Earth. Examples of evidence for the model include observations of the masses and lifetimes of other stars, as well as the ways that the Sun's radiation varies due to sudden solar flares ("space weather"), the 11-year sunspot cycle, and non-cyclic variations over centuries. Assessment Boundary: Assessment does not include details of the atomic and subatomic processes involved with the Sun's nuclear fusion.	**ESS1.A. The Universe and Its Stars** The star called the Sun is changing and will burn out over a life span of approximately 10 billion years. **PS3.D. Energy in Chemical Processes and Everyday Life** Nuclear fusion processes in the center of the Sun release the energy that ultimately reaches Earth as radiation.
HS-ESS1-2. Construct an explanation of the big bang theory based on astronomical evidence of light spectra, motion of distant galaxies, and composition of matter in the universe. **Clarification Statement:** Emphasis is on the astronomical evidence of the red shift of light from galaxies as an indication that the universe is currently expanding, the cosmic microwave background as the remnant radiation from the big bang, and the observed composition of ordinary matter of the universe, primarily found in stars and interstellar gases (from the spectra of electromagnetic radiation from stars), which matches that predicted by the big bang theory (3/4 hydrogen and 1/4 helium).	**PS4.B. Electromagnetic Radiation** Atoms of each element emit and absorb characteristic frequencies of light. These characteristics allow identification of the presence of an element, even in microscopic quantities. **ESS1.A. The Universe and Its Stars** The study of stars' light spectra and brightness is used to identify compositional elements of stars, their movements, and their distances from Earth. (HS-ESS1-3) The big bang theory is supported by observations of distant galaxies receding from our own, of the measured composition of stars and non-stellar gases, and of the maps of spectra of the primordial radiation (cosmic microwave background) that still fills the universe. Other than the hydrogen and helium formed at the time of the big bang, nuclear fusion within stars produces all atomic nuclei lighter than and including iron, and the process releases electromagnetic energy. Heavier elements are produced when certain massive stars achieve a supernova stage and explode. (HS-ESS1-3)
HS-ESS1-3. Communicate scientific ideas about the way stars, over their life cycle, produce elements. **Clarification Statement:** Emphasis is on the way nucleosynthesis, and therefore the different elements created, varies as a function of the mass of a star and the stage of its lifetime. **Assessment Boundary:** Details of the many different nucleosynthesis pathways for stars of differing masses are not assessed.	**ESS1.A. The Universe and Its Stars** The study of stars' light spectra and brightness is used to identify compositional elements of stars, their movements, and their distances from Earth. (HS-ESS1-2) Other than the hydrogen and helium formed at the time of the big bang, nuclear fusion within stars produces all atomic nuclei lighter than and including iron, and the process releases electromagnetic energy. Heavier elements are produced when certain massive stars achieve a supernova stage and explode. (HS-ESS1-2)

Performance Expectations and Disciplinary Core Ideas for Earth and Space Science (*continued*)

Performance Expectations (PEs)	Disciplinary Core Ideas (DCIs)
HS-ESS1-4. Use mathematical or computational representations to predict the motion of orbiting objects in the solar system. **Clarification Statement:** Emphasis is on Newtonian gravitational laws governing orbital motions, which apply to human-made satellites as well as planets and Moons. **Assessment Boundary:** Mathematical representations for the gravitational attraction of bodies and Kepler's laws of orbital motions should not deal with more than two bodies, nor involve calculus.	**ESS1.B. Earth and the Solar System** Kepler's laws describe common features of the motions of orbiting objects, including their elliptical paths around the Sun. Orbits may change due to the gravitational effects from, or collisions with, other objects in the solar system.
HS-ESS1-5. Evaluate evidence of the past and current movements of continental and oceanic crust and the theory of plate tectonics to explain the ages of crustal rocks. **Clarification Statement:** Emphasis is on the ability of plate tectonics to explain the ages of crustal rocks. Examples include evidence of the ages of oceanic crust increasing with distance from mid-ocean ridges (a result of plate spreading) and the ages of North American continental crust increasing with distance away from a central ancient core (a result of past plate interactions).	**ESS2.B. Plate Tectonics and Large-Scale System Interactions** Plate tectonics is the unifying theory that explains the past and current movements of the rocks at Earth's surface and provides a framework for understanding its geologic history. Plate movements are responsible for most continental and ocean-floor features and for the distribution of most rocks and minerals within Earth's crust. (HS-ESS2-1) **ESS1.C. The History of Planet Earth** Continental rocks, which can be older than 4 billion years, are generally much older than the rocks of the ocean floor, which are less than 200 million years old. **PS1.C. Nuclear Processes** Spontaneous radioactive decays follow a characteristic exponential decay law. Nuclear lifetimes allow radiometric dating to be used to determine the ages of rocks and other materials. (HS-ESS1-6)
HS-ESS1-6. Apply scientific reasoning and evidence from ancient Earth materials, meteorites, and other planetary surfaces to construct an account of Earth's formation and early history. **Clarification Statement:** Emphasis is on using available evidence within the solar system to reconstruct the early history of Earth, which formed along with the rest of the solar system 4.6 billion years ago. Examples of evidence include the absolute ages of ancient materials (obtained by radiometric dating of meteorites, Moon rocks, and Earth's oldest minerals), the sizes and compositions of solar system objects, and the impact cratering record of planetary surfaces.	**PS1.C. Nuclear Processes** Spontaneous radioactive decays follow a characteristic exponential decay law. Nuclear lifetimes allow radiometric dating to be used to determine the ages of rocks and other materials. (HS-ESS1-6) **ESS1.C. The History of Planet Earth** Although active geologic processes, such as plate tectonics and erosion, have destroyed or altered most of the very early rock record on Earth, other objects in the solar system, such as lunar rocks, asteroids, and meteorites, have changed little over billions of years. Studying these objects can provide information about Earth's formation and early history.

Performance Expectations and Disciplinary Core Ideas for Earth and Space Science (*continued*)

Performance Expectations (PEs)	Disciplinary Core Ideas (DCIs)
HS-ESS2-1. Develop a model to illustrate how Earth's internal and surface processes operate at different spatial and temporal scales to form continental and ocean-floor features. **Clarification Statement:** Emphasis is on how the appearance of land features (such as mountains, valleys, and plateaus) and seafloor features (such as trenches, ridges, and seamounts) are a result of both constructive forces (such as volcanism, tectonic uplift, and orogeny) and destructive mechanisms (such as weathering, mass wasting, and coastal erosion). **Assessment Boundary:** Assessment does not include memorization of the details of the formation of specific geographic features of Earth's surface.	**ESS2.A. Earth Materials and Systems** Earth's systems, being dynamic and interacting, cause feedback effects that can increase or decrease the original changes. (HS-ESS2-2) **ESS2.B. Plate Tectonics and Large-Scale System Interactions** Plate tectonics is the unifying theory that explains the past and current movements of the rocks at Earth's surface and provides a framework for understanding its geologic history. Plate movements are responsible for most continental and ocean-floor features and for the distribution of most rocks and minerals within Earth's crust. (HS-ESS1-5)
HS-ESS2-2. Analyze geoscience data to make the claim that one change to Earth's surface can create feedbacks that cause changes to other Earth systems. **Clarification Statement:** Examples should include climate feedbacks, such as how an increase in greenhouse gases causes a rise in global temperatures that melts glacial ice, which reduces the amount of sunlight reflected from Earth's surface, increasing surface temperatures and further reducing the amount of ice. Examples could also be taken from other system interactions, such as how the loss of ground vegetation causes an increase in water runoff and soil erosion; how dammed rivers increase groundwater recharge, decrease sediment transport, and increase coastal erosion; or how the loss of wetlands causes a decrease in local humidity that further reduces the wetland extent.	**ESS2.D. Weather and Climate** The foundation for Earth's global climate systems is the electromagnetic radiation from the Sun, as well as its reflection, absorption, storage, and redistribution among the atmosphere, ocean, and land systems, and this energy's re-radiation into space. (HS-ESS2-4) **ESS2.A. Earth Materials and Systems** Earth's systems, being dynamic and interacting, cause feedback effects that can increase or decrease the original changes. (HS-ESS2-1)
HS-ESS2-3. Develop a model based on evidence of Earth's interior to describe the cycling of matter by thermal convection. **Clarification Statement:** Emphasis is on both a one-dimensional model of Earth, with radial layers determined by density, and a three-dimensional model, which is controlled by mantle convection and the resulting plate tectonics. Examples of evidence include maps of Earth's three-dimensional structure obtained from seismic waves, records of the rate of change of Earth's magnetic field (as constraints on convection in the outer core), and identification of the composition of Earth's layers from high-pressure laboratory experiments.	**ESS2.A. Earth Materials and Systems** Evidence from deep probes and seismic waves, reconstructions of historical changes in Earth's surface and its magnetic field, and an understanding of physical and chemical processes lead to a model of Earth with a hot but solid inner core, a liquid outer core, and a solid mantle and crust. Motions of the mantle and its plates occur primarily through thermal convection, which involves the cycling of matter due to the outward flow of energy from Earth's interior and gravitational movement of denser materials toward the interior. **ESS2.B. Plate Tectonics and Large-Scale System Interactions** The radioactive decay of unstable isotopes continually generates new energy within Earth's crust and mantle, providing the primary source of the heat that drives mantle convection. Plate tectonics can be viewed as the surface expression of mantle convection.

Performance Expectations and Disciplinary Core Ideas for Earth and Space Science (*continued*)

Performance Expectations (PEs)	Disciplinary Core Ideas (DCIs)
HS-ESS2-4. Use a model to describe how variations in the flow of energy into and out of Earth's systems result in changes in climate. **Clarification Statement:** Examples of the causes of climate change differ by time scale; over one to ten years: large volcanic eruption, ocean circulation; tens to hundreds of years: changes in human activity, ocean circulation, solar output; tens to hundreds of thousands of years: changes to Earth's orbit and the orientation of its axis; and tens to hundreds of millions of years: long-term changes in atmospheric composition. **Assessment Boundary:** Assessment of the results of changes in climate is limited to changes in surface temperatures, precipitation patterns, glacial ice volumes, sea levels, and biosphere distribution.	**ESS2.A. Earth Materials and Systems** The geologic record shows that changes to global and regional climate can be caused by interactions among changes in the Sun's energy output or Earth's orbit, tectonic events, ocean circulation, volcanic activity, glaciers, vegetation, and human activities. These changes can occur on a variety of time scales from sudden (e.g., volcanic ash clouds) to intermediate (ice ages) to very long-term tectonic cycles. **ESS1.B. Earth and the Solar System** Cyclical changes in the shape of Earth's orbit around the Sun, together with changes in the tilt of the planet's axis of rotation, both occurring over hundreds of thousands of years, have altered the intensity and distribution of sunlight falling on the Earth. These phenomena cause a cycle of ice ages and other gradual climate changes. **ESS2.D. Weather and Climate** The foundation for Earth's global climate systems is the electromagnetic radiation from the Sun, as well as its reflection, absorption, storage, and redistribution among the atmosphere, ocean, and land systems, and this energy's re-radiation into space. (HS-ESS2-2) Changes in the atmosphere due to human activity have increased carbon dioxide concentrations and thus affect climate. (HS-ESS2-6)
HS-ESS2-5. Plan and conduct an investigation of the properties of water and its effects on Earth materials and surface processes. **Clarification Statement:** Emphasis is on mechanical and chemical investigations with water and a variety of solid materials to provide the evidence for connections between the hydrologic cycle and system interactions commonly known as the rock cycle. Examples of mechanical investigations include stream transportation and deposition using a stream table, erosion using variations in soil moisture content, or frost wedging by the expansion of water as it freezes. Examples of chemical investigations include chemical weathering and recrystallization (by testing the solubility of different materials) or melt generation (by examining how water lowers the melting temperature of most solids).	**ESS2.C. The Roles of Water in Earth's Surface Processes** The abundance of liquid water on Earth's surface and its unique combination of physical and chemical properties are central to the planet's dynamics. These properties include water's exceptional capacity to absorb, store, and release large amounts of energy, transmit sunlight, expand upon freezing, dissolve and transport materials, and lower the viscosities and melting points of rocks.
HS-ESS2-6. Develop a quantitative model to describe the cycling of carbon among the hydrosphere, atmosphere, geosphere, and biosphere. **Clarification Statement:** Emphasis is on modeling biogeochemical cycles that include the cycling of carbon through the ocean, atmosphere, soil, and biosphere (including humans), providing the foundation for living organisms.	**ESS2.D. Weather and Climate** Gradual atmospheric changes were due to plants and other organisms that captured carbon dioxide and released oxygen. (HS-ESS2-7) Changes in the atmosphere due to human activity have increased carbon dioxide concentrations and thus affect climate. (HS-ESS2-4)

Performance Expectations and Disciplinary Core Ideas for Earth and Space Science (*continued*)

Performance Expectations (PEs)	Disciplinary Core Ideas (DCIs)
HS-ESS2-7. Construct an argument based on evidence about the simultaneous co-evolution of Earth's systems and life on Earth. **Clarification Statement:** Emphasis is on the dynamic causes, effects, and feedbacks between the biosphere and Earth's other systems, whereby geoscience factors control the evolution of life, which in turn continuously alters Earth's surface. Examples include how photosynthetic life altered the atmosphere through the production of oxygen, which in turn increased weathering rates and allowed for the evolution of animal life; how microbial life on land increased the formation of soil, which in turn allowed for the evolution of land plants; or how the evolution of corals created reefs that altered patterns of erosion and deposition along coastlines and provided habitats for the evolution of new life forms. **Assessment Boundary:** Assessment does not include a comprehensive understanding of the mechanisms of how the biosphere interacts with all of Earth's other systems.	**ESS2.D. Weather and Climate** Gradual atmospheric changes were due to plants and other organisms that captured carbon dioxide and released oxygen. (HS-ESS2-6) **ESS2.E. Biogeology** The many dynamic and delicate feedbacks between the biosphere and other Earth systems cause a continual co-evolution of Earth's surface and the life that exists on it.
HS-ESS3-1. Construct an explanation based on evidence for how the availability of natural resources, occurrence of natural hazards, and changes in climate have influenced human activity. **Clarification Statement:** Examples of key natural resources include access to fresh water (such as rivers, lakes, and groundwater), regions of fertile soils such as river deltas, and high concentrations of minerals and fossil fuels. Examples of natural hazards can be from interior processes (such as volcanic eruptions and earthquakes), surface processes (such as tsunamis, mass wasting, and soil erosion), and severe weather (such as hurricanes, floods, and droughts). Examples of the results of changes in climate that can affect populations or drive mass migrations include changes to sea level, regional patterns of temperature and precipitation, and the types of crops and livestock that can be raised.	**ESS3.A. Natural Resources** Resource availability has guided the development of human society. **ESS3.B. Natural Hazards** Natural hazards and other geologic events have shaped the course of human history; they have significantly altered the sizes of human populations and have driven human migrations.
HS-ESS3-2. Evaluate competing design solutions for developing, managing, and utilizing energy and mineral resources based on cost-benefit ratios. **Clarification Statement:** Emphasis is on the conservation, recycling, and reuse of resources (such as minerals and metals) where possible, and on minimizing impacts where it is not. Examples include developing best practices for agricultural soil use, mining (for coal, tar sands, and oil shales), and pumping (for petroleum and natural gas). Science knowledge indicates what can happen in natural systems—not what should happen.	**ESS3.A. Natural Resources** All forms of energy production and other resource extraction have associated economic, social, environmental, and geopolitical costs and risks as well as benefits. New technologies and social regulations can change the balance of these factors. **ETS1.B. Developing Possible Solutions** When evaluating solutions it is important to take into account a range of constraints including cost, safety, reliability, and aesthetics and to consider social, cultural, and environmental impacts. (HS-ESS3-4), (HS-ETS1-3), (HS-LS2-7), (HS-LS4-6)

Performance Expectations and Disciplinary Core Ideas for Earth and Space Science (*continued*)

Performance Expectations (PEs)	Disciplinary Core Ideas (DCIs)
HS-ESS3-3. Create a computational simulation to illustrate the relationships among management of natural resources, the sustainability of human populations, and biodiversity. **Clarification Statement:** Examples of factors that affect the management of natural resources include costs of resource extraction and waste management, per capita consumption, and the development of new technologies. Examples of factors that affect human sustainability include agricultural efficiency, levels of conservation, and urban planning. **Assessment Boundary:** Assessment for computational simulations is limited to using provided multi-parameter programs or constructing simplified spreadsheet calculations.	**ESS3.C. Human Impacts on Earth Systems** The sustainability of human societies and the biodiversity that supports them requires responsible management of natural resources.
HS-ESS3-4. Evaluate or refine a technological solution that reduces impacts of human activities on natural systems. **Clarification Statement:** Examples of data on the impacts of human activities could include the quantities and types of pollutants released, changes to biomass and species diversity, or areal changes in land surface use (such as for urban development, agriculture and livestock, or surface mining). Examples for limiting future impacts could range from local efforts (such as reducing, reusing, and recycling resources) to large-scale geoengineering design solutions (such as altering global temperatures by making large changes to the atmosphere or ocean).	**ESS3.C. Human Impacts on Earth Systems** Scientists and engineers can make major contributions by developing technologies that produce less pollution and waste and that preclude ecosystem degradation. **ETS1.B. Developing Possible Solutions** When evaluating solutions it is important to take into account a range of constraints including cost, safety, reliability, and aesthetics and to consider social, cultural, and environmental impacts.
HS-ESS3-5. Analyze geoscience data and the results from global climate models to make an evidence-based forecast of the current rate of global or regional climate change and associated future impacts to Earth systems. **Clarification Statement:** Examples of evidence, for both data and climate model outputs, are for climate changes (such as precipitation and temperature) and their associated impacts (such as on sea level, glacial ice volumes, or atmosphere and ocean composition). **Assessment Boundary:** Assessment is limited to one example of a climate change and its associated impacts.	**ESS3.D. Global Climate Change** Though the magnitudes of human impacts are greater than they have ever been, so too are human abilities to model, predict, and manage current and future impacts.
HS-ESS3-6. Use a computational representation to illustrate the relationships among Earth systems and how those relationships are being modified due to human activity. **Clarification Statement:** Examples of Earth systems to be considered are the hydrosphere, atmosphere, cryosphere, geosphere, and/or biosphere. An example of the far-reaching impacts from a human activity is how an increase in atmospheric carbon dioxide results in an increase in photosynthetic biomass on land and an increase in ocean acidification, with resulting impacts on sea organism health and marine populations. **Assessment Boundary:** Assessment does not include running computational representations but is limited to using the published results of scientific computational models.	**ESS2.D. Weather and Climate** Current models predict that, although future regional climate changes will be complex and varied, average global temperatures will continue to rise. The outcomes predicted by global climate models strongly depend on the amounts of human-generated greenhouse gases added to the atmosphere each year and by the ways in which these gases are absorbed by the ocean and biosphere. **ESS3.D. Global Climate Change** Through computer simulations and other studies, important discoveries are still being made about how the ocean, the atmosphere, and the biosphere interact and are modified in response to human activities.

Performance Expectations and Disciplinary Core Ideas for Engineering Design

Performance Expectations (PEs)	Disciplinary Core Ideas (DCIs)
HS-ETS1-1. Analyze a major global challenge to specify qualitative and quantitative criteria and constraints for solutions that account for societal needs and wants.	**ETS1.A. Defining and Delimiting Engineering Problems** Criteria and constraints also include satisfying any requirements set by society, such as taking issues of risk mitigation into account, and they should be quantified to the extent possible and stated in such a way that one can tell if a given design meets them. (HS-PS2-3) Humanity faces major global challenges today, such as the need for supplies of clean water and food or for energy sources that minimize pollution, which can be addressed through engineering. These global challenges also may have manifestations in local communities.
HS-ETS1-2. Design a solution to a complex real-world problem by breaking it down into smaller, more manageable problems that can be solved through engineering.	**ETS1.C. Optimizing the Design Solution** Criteria may need to be broken down into simpler ones that can be approached systematically, and decisions about the priority of certain criteria over others (trade-offs) may be needed. (HS-PS1-6), (HS-PS2-3)
HS-ETS1-3. Evaluate a solution to a complex real-world problem based on prioritized criteria and trade-offs that account for a range of constraints, including cost, safety, reliability, and aesthetics as well as possible social, cultural, and environmental impacts.	**ETS1.B. Developing Possible Solutions** When evaluating solutions it is important to take into account a range of constraints including cost, safety, reliability, and aesthetics and to consider social, cultural, and environmental impacts. (HS-LS4-6)
HS-ETS1-4. Use a computer simulation to model the impact of proposed solutions to a complex real-world problem with numerous criteria and constraints on interactions within and between systems relevant to the problem.	**ETS1.B. Developing Possible Solutions** Both physical models and computers can be used in various ways to aid in the engineering design process. Computers are useful for a variety of purposes, such as running simulations to test different ways of solving a problem or to see which one is most efficient or economical, and in making a persuasive presentation to a client about how a given design will meet his or her needs. (HS-LS4-6)